Magnetical Investigations

Volume 1

William Scoresby

CAMBRIDGE UNIVERSITY PRESS

Cambridge, New York, Melbourne, Madrid, Cape Town,
Singapore, São Paolo, Delhi, Mexico City

Published in the United States of America by Cambridge University Press, New York

www.cambridge.org
Information on this title: www.cambridge.org/9781108052627

© in this compilation Cambridge University Press 2012

This edition first published 1839
This digitally printed version 2012

ISBN 978-1-108-05262-7 Paperback

CAMBRIDGE LIBRARY COLLECTION

Books of enduring scholarly value

Technology

The focus of this series is engineering, broadly construed. It covers technological innovation from a range of periods and cultures, but centres on the technological achievements of the industrial era in the West, particularly in the nineteenth century, as understood by their contemporaries. Infrastructure is one major focus, covering the building of railways and canals, bridges and tunnels, land drainage, the laying of submarine cables, and the construction of docks and lighthouses. Other key topics include developments in industrial and manufacturing fields such as mining technology, the production of iron and steel, the use of steam power, and chemical processes such as photography and textile dyes.

Magnetical Investigations

Published between 1839 and 1852, this two-volume work records the contribution of William Scoresby (1789–1857) to magnetic science, a field he considered one of 'grandeur'. The result of laborious investigations into magnetism and (with James Prescott Joule) electromagnetism, Scoresby's work was particularly concerned with improving the accuracy of ships' compasses. A whaler, scientist and clergyman, he epitomised the contribution which could be made to exploration and science by provincial merchant mariners – men often less celebrated than their counterparts in the Royal Navy or in metropolitan learned societies. In addition to his pioneering work on magnetic science, Scoresby furthered knowledge of Arctic meteorology, oceanography and geography. Volume 1 considers the magnetism of steel and suggests ways to determine its quality and hardness.

Cambridge University Press has long been a pioneer in the reissuing of out-of-print titles from its own backlist, producing digital reprints of books that are still sought after by scholars and students but could not be reprinted economically using traditional technology. The Cambridge Library Collection extends this activity to a wider range of books which are still of importance to researchers and professionals, either for the source material they contain, or as landmarks in the history of their academic discipline.

Drawing from the world-renowned collections in the Cambridge University Library and other partner libraries, and guided by the advice of experts in each subject area, Cambridge University Press is using state-of-the-art scanning machines in its own Printing House to capture the content of each book selected for inclusion. The files are processed to give a consistently clear, crisp image, and the books finished to the high quality standard for which the Press is recognised around the world. The latest print-on-demand technology ensures that the books will remain available indefinitely, and that orders for single or multiple copies can quickly be supplied.

The Cambridge Library Collection brings back to life books of enduring scholarly value (including out-of-copyright works originally issued by other publishers) across a wide range of disciplines in the humanities and social sciences and in science and technology.

MAGNETICAL

INVESTIGATIONS.

BY THE

REV. WILLIAM SCORESBY, B.D.

FELLOW OF THE ROYAL SOCIETIES OF LONDON AND EDINBURGH;
CORRESPONDING MEMBER OF THE INSTITUTE OF FRANCE,
ETC. ETC.

PART I.

COMPRISING INVESTIGATIONS ON THE PRINCIPLES
AFFECTING THE CAPACITY AND RETENTIVENESS OF STEEL
FOR THE MAGNETIC CONDITION;
WITH THE DEVELOPMENT OF PROCESSES FOR DETERMINING
THE QUALITY AND DEGREE OF HARDNESS OF STEEL.

LONDON:

LONGMAN, ORME, BROWN, GREEN, AND LONGMANS,

PATERNOSTER-ROW.

M.DCCC.XXXIX.

TO

THE REV. WILLIAM WHEWELL, M.A.

PROFESSOR OF CASUISTRY IN THE UNIVERSITY OF CAMBRIDGE,

THESE

MAGNETICAL INVESTIGATIONS,

AS A TESTIMONY OF RESPECT FOR HIS HIGH TALENTS,
AND AS AN EXPRESSION OF PERSONAL ESTEEM,

ARE INSCRIBED BY

THE AUTHOR.

Exeter, 11th March, 1839.

CONTENTS.

PART I.

CONTENTS.

CHAPTER IV.

CHAPTER V.

CHAPTER VI.

CHAPTER VII.

CHAPTER VIII.

Plate 1.

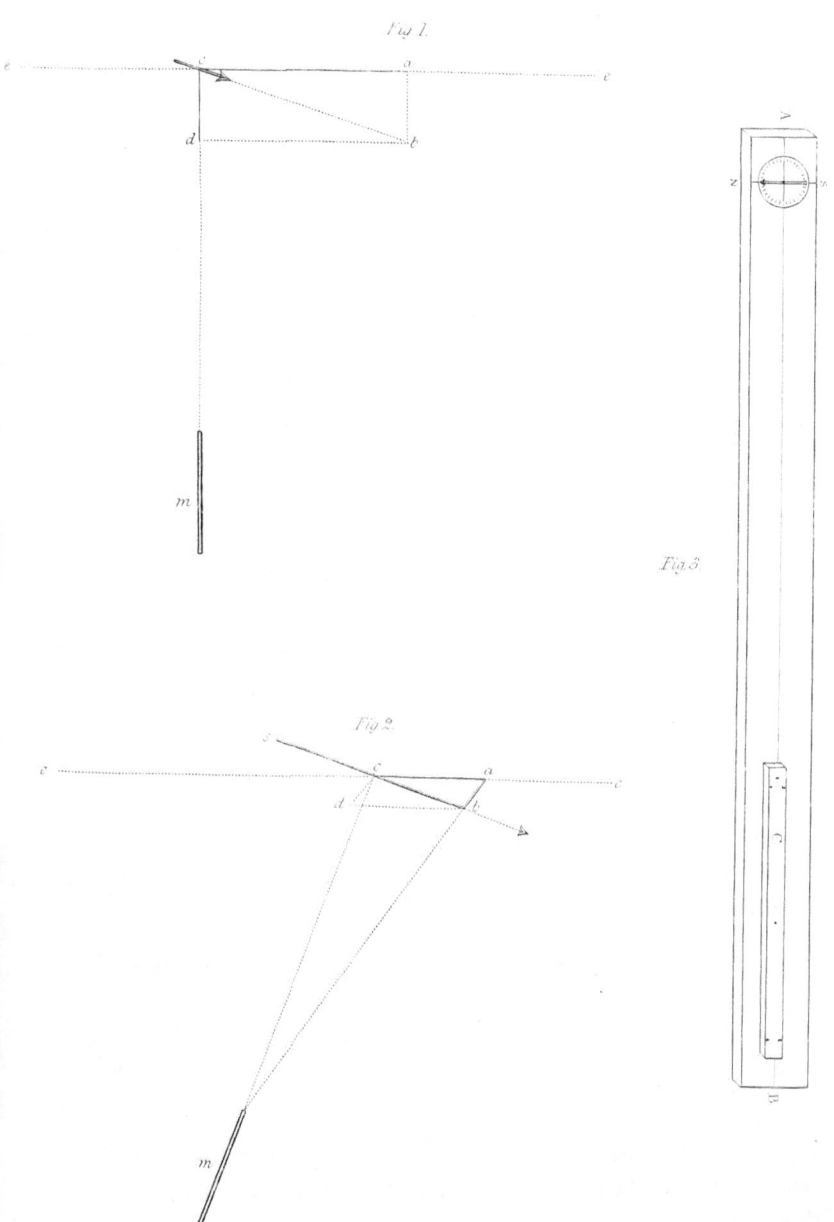

Fig. 1.

Fig. 2.

Fig. 3.

London, Longman & C.º Paternoster Row

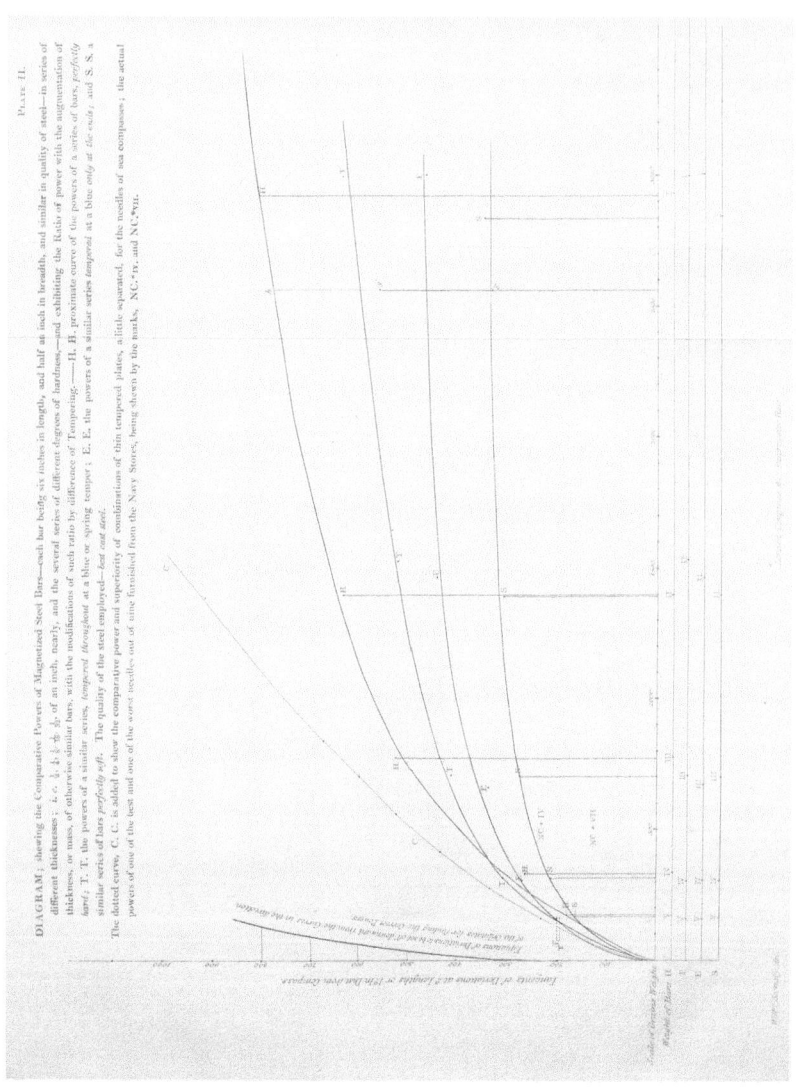

DIAGRAM; shewing the Comparative Powers of Magnetized Steel Bars—each bar being six inches in length, and half an inch in breadth, and similar in quality of steel—in series of different thicknesses; i. e. $\frac{1}{8}$, $\frac{1}{4}$, $\frac{3}{8}$, $\frac{1}{2}$ of an inch, nearly, and the several series of different degrees of hardness,—and exhibiting the Ratio of power with the augmentation of thickness, or mass, of otherwise similar bars, with the modification of such ratio by difference of Tempering. —— H, H, proximate curve of the powers of a series of bars, *perfectly hard*; T, T, the powers of a similar series, *tempered throughout* at a blue or spring temper; E, E, the powers of a similar series tempered at a blue only at *the ends*, and S, S, a similar series of bars *perfectly soft*. The quality of the steel employed—*best cast steel*.

The dotted curve, C, C, is added to shew the comparative power and superiority of combinations of thin tempered plates, a little separated, for the needles of sea compasses; the actual powers of one of the best and one of the worst needles out of nine furnished from the Navy Stores, being shewn by the marks, NC.⁴ IV, and NC.⁹ VII.

The material originally positioned here is too large for reproduction in this reissue. A PDF can be downloaded from the web address given on page iv of this book, by clicking on 'Resources Available'.

MAGNETICAL INVESTIGATIONS.

PART I.

CHAPTER I.

INTRODUCTORY OBSERVATIONS.

IT had long been conjectured that a most inti-
mate connexion, if not identity of nature, might
probably exist among some of the more subtle
and mysterious agents, or principles, of different denominations, and, apparently, of different
characteristics, which universally pervade the
region in and about this earth.

Modern discoveries in electro-magnetism—
with the cognate relations which these have developed as existing in other principles of natural
bodies—have gone far to verify these anticipations; and at the same time to yield so much

B

additional knowledge of the constitution of the physical system of our planet, as to give a new, an interesting, and a prominent importance to electrical and magnetical science.

To the time of Dr. Gilbert of Colchester, magnetism was only known as a mysterious virtue, existing in, and peculiar to, the loadstone, or to ferruginous substances which had been touched by this extraordinary mineral, from which certain qualities of attraction and direction were derived. But this eminent individual discerned—as he has left on record in his "Physiologia Nova, seu Tractatus de Magnete et Corporibus Magneticis;" published in the year 1600 — that the phenomenon of the meridional adjustment of the magnetic needle was not owing to any mystical virtue exercised out of the course of natural principles, but a mere result of the directive action of the earth; which he truly considered as the controlling agent, by reason of its being in its matter and constitution magnetic.

So long as magnetism was known only as a separate or simple principle, the philosophic ideas of Dr. Gilbert were never materially advanced; but on the discoveries of Professor Oersted, whereby the long-suspected connexion betwixt electricity and magnetism was established, an amazing enlargement was at once yielded to magnetical knowledge, and a corresponding impulse given to magnetical research.

The effect has been to give to a science, formerly considered as comparatively of an inferior class, a grandeur of consideration ; placing it at once amongst those mighty principles which Infinite Wisdom has appointed, and Infinite Power ordained, as essential elements or agencies in the physical constitution of the world. For inasmuch as an inseparable connexion has been established betwixt electricity and magnetism,—these, under various forms of development, reciprocally developing each other—it necessarily follows, that, to whatever extent in creation electricity operates, to the same extent the magnetic principle must reach. And inasmuch as heat, light, and chemical action, are each, more or less, developers of one form or other of the electro-magneto principle, the analogies of science would lead us to infer, that magnetism is co-extensive with these other agencies, throughout their range of operation.

Hence, there is little doubt but the principle, which, a few years ago, was known only as a director of the compass needle, as to its utility, and as little more than a curiosity in science—is one of the mighty energies by which, instrumentally, the works of the great Creator are regulated ; one of those subtle powers which He hath ordained as his servants, "fulfilling his word;" and whereby "the sweet influences" of the whole system of the universe are bound together, controlled, and upheld.

Thus the subject of magnetism becomes of the highest consideration;—in science, as to its mightiness and extent of operation—and in natural theology, as calculated to connect the researches of human intelligence, with Him who hath created these wonders; to elevate the feelings of reverence and adoration in the devotional mind, and to proclaim more clearly, in proportion as these invisible things are understood, "His eternal power and Godhead."

It is not my object, however, in this publication, to carry out those views to which the more enlarged consideration of magnetism, as a science, might be advantageously applied; but that, in thus shewing something of the importance of the subject as a science, I may solicit, for the contributions which are here offered to it, such reasonable consideration as, in this connexion, they may fairly claim.

To the subject of *magnetism*, my attention has for a series of years been more or less directed; latterly with the view, particularly, of producing more powerful instruments for the determination of delicate variations in, and the actual condition of, the earth's magnetism—a subject which, from its greatly increased importance, is now engaging attention in some of the principal observatories in Europe.

In contemplating such improvement in instruments dependent for their adjustment on the earth's magnetism, the grand desideratum would obviously be, the attainment of increased energy, or directive power, in magnetic needles, or bars of any given length and mass. And, that the attainment of such increased energy was a promising field of inquiry, I was satisfied, from the mere consideration of the surprising superiority in power of electro-magnets over permanent artificial magnets;—strongly indicating the existence of a far greater capacity for magnetism, in the masses of steel commonly employed for magnets, than we have hitherto been able to develope, or, if developed, to retain.

Whilst this consideration yielded every encouragement in the inquiry, an experimental fact, in regard to the proportional power of magnets of *unequal thickness*, suggested that guidance, in pursuing the inquiry, which not only led (as will, I trust, subsequently appear) to a successful result in regard to the object particularly specified, but gave rise to investigations extending beyond my original design, and eminently calculated, I conceive, for the improvement of sea-compasses—hitherto so very defective—as well as of artificial magnets, and magnetic apparatus generally. The fact referred to was this :—

When examining, many years ago, the directive power of various artificial bar-magnets, for the purpose of determining the practicability of ascertaining the thickness of rocks, etc. in tunnelling and mining, by the method of deviations communicated to the Royal Society in 1831,— the idea occurred to me, that, if the bars ordinarily employed for compass-needles, etc., were divided into laminæ; or if, in other words, they were made up of thin plates to the extent of the masses of the bars commonly in use, a greater degree of energy would be obtained. Experiment fully justified this opinion.

But previous to the application of the principle to instruments directed by the earth's magnetism, I had made trial of a combination of laminæ of *thoroughly tempered steel*, for the construction of a small compound magnet. The substance made use of was watch-spring, of which fourteen pieces, of two inches in length, were combined, after being magnetized, and formed a small magnet, weighing, in amount of steel, about one hundred grains. From a want of knowledge, at that time, of the best mode of magnetizing thin plates, the power obtained was

much less than was expected ; but when the same
little instrument was subsequently magnetized,
in the mass, by the process hereafter described,
its efficiency became very striking, — the power
being found to be such as to lift, by one pole, a
polished nail of about 800 grains, or eight times
its own weight.

A trial-apparatus, of the nature of a *variation-
needle*, on the same principle—improved, however, for this purpose, by the *separation* of the
plates—was constructed in the year 1836, which
was exhibited to the " British Association for
the promotion of Science," the same year. But
being without any *precise* knowledge of the laws
of combination in magnetized plates, or even of
the actual power of this new instrument, though
obviously great,—a mere general idea only of
its relative superiority could then be obtained.

Since that time I have investigated, in an
elaborate series of experiments, and with a
somewhat expensive variety of apparatus, the
principles of the construction adopted, so as
satisfactorily to prove, I conceive, the decided
advantage of that construction for sea, and other,
compasses, and to apply the principle to various
practical purposes in magnetics.

In the original " variation-compass" just referred to, the plates, as I have intimated, were not
placed in immediate contact, but separated by thin

slips of wood or cardboard ; for I had previously
found, when combining magnets for other purposes, that a material loss of power in the individual intensities of the bars, was, in all cases,
occasioned by combination ; but that, when the
combination was not made in contact, an inferior
deterioration took place.

The subjects to which I was primarily guided
by these preliminary considerations and results,
extended to the following particulars : — The
effect of the division in various directions of the
mass of steel, on its magnetic capabilities,—the
law of combination of magnetized steel plates in
contact—the law of combination when the plates
are separated by limited spaces—the effect of
temper or degree of hardness, and the degree of
permanency of the power in combinations of
magnetized steel plates.   A few of the most
important results obtained from these investigations, were forwarded to the Institute of France,
in February 1838.

These investigations proved sufficient to shew
that the idea entertained in the outset,—of the
practicability of producing, by means of a combination of magnetized steel plates, more powerful apparatus than had hitherto been in use, for
experiment and observation in magnetical science, and for practical purposes in magnetics
generally,—was not incorrectly founded.

But the most important practical applications

of these principles—with the results of several new and distinct investigations — yet remain behind. The description of these, in the first instance, is the object of the present publication; and it is hoped that the results will be found to develope some new and some improved principles of construction, applicable both to instruments designed to be directed by the earth's magnetism, and to the improvement of apparatus in which a permanent and concentrate energy are requisite; together with a useful application of some of the laws developed in these or previous investigations, to the *testing* of the quality of steel, and the determining of the degree of its hardness —purposes of the highest importance in the construction of delicate instruments used in the arts, or by professional men.

The several results and applications of these recent personal researches, may be conveniently classed under separate heads, belonging to the development of principles, or of different practical processes, in magnetism. Some of these will, no doubt, be resolvable, in certain particulars, into principles or methods heretofore known; but in all, it is presumed, something peculiar, as to decisiveness of the results bearing on principles, or as to convenience of adaptation, or efficacy of manipulation, or improvement in construction, in regard to the practical subjects, will be found.

CHAPTER II.

ALTHOUGH the processes used in these investigations for developing the magnetic condition, are resolvable into methods already well known; yet, as by a vast number of experiments and trials on many hundreds of plates and bars, comprising great variety of size, form, quality and degree of hardness, they have been found to be decidedly more efficient, convenient, and equable in their results, than any other of the processes usually practised, with most of which they have been very particularly compared,—it may be useful to others who may make comparative experiments, or who may wish to apply the principles developed to the practical purposes for which they are calculated, to describe the processes particularly.

These processes—consisting only of two varieties of manipulation—were found to be most effectively applicable for the magnetizing of every kind of steel bar which I had occasion to employ in a most extensive series of investigations : the one process for developing the maximum energy in very large bars, when in pairs; the other in small single bars, as well as in very thin plates, of every variety of magnitude, and every degree of hardness.

For the magnetizing of *large bars* in pairs,— I have found no process so efficient, or so convenient and rapid in operation, as that of Æpinus —the developing power consisting of a large compound horse-shoe magnet. But this method I find modified to much advantage, (as originally suggested to me by a practical magnetician) and with a very striking influence on the ultimate power, especially in large massive hard bars, by passing the horse-shoe magnet employed in the process *round the whole parallelogram of steel bars and iron conductors, in the same direction*, (terminating at the middle of one of the bars) instead of limiting the manipulations to the extent of the steel bars separately from end to end.

In effecting all that is needful in a pair of bars tempered from end to end; 2 feet in length, 1·5 inches broad, and 0·6 inch in thickness,— two circuits only of the parallelogram, on each side of the bars, by the large magnet, were

necessary. It is important however, that the
passage of the magnet should be made smoothly
and without hitchings; the effecting of which
with any degree of certainty, requires the surfaces of the bars to be slightly oiled. Though
the interposition of oil may hinder in some
measure the closeness of the contact, I have
not found that circumstance—where a powerful
horse-shoe magnet has been employed—of any
consequence; but on the whole, a benefit, even
in magnetizing bars so large as three feet in
length, and two and a half inches in width.
But the modification of the process, in forming
the circuit of the parallelogram, yielded, on
trial, about one-seventh additional power in the
heavy two-feet bars, above the maximum energy
I was able to develope by the exact method of
Æpinus. This process, so modified, is distinguished in the following descriptions, as the
method ÆS.

For the magnetizing of *thin plates* of all descriptions and dimensions, up to the measure in
length and breadth of the magnets employed in
the operation; the process ultimately adopted—
being a process most simple, rapid and effective
—was a modification of that of Dr. Knight. For,
by a change in the arrangement, according to
the description usually given of Dr. Knight's
method, a most important practical advantage

is gained. This change mainly consists in placing the plate or bar to be magnetized *above*, instead of *beneath* the magnets employed in the operation; by which very great facility is given for the performance of the requisite manipulations, and for the maximum development of the magnetic condition.

A pair of powerful bar magnets (single bars tempered, or made hard, throughout) of equal length and breadth, at least, with those dimensions of the plates to be magnetized, are placed in a straight line, with their opposite poles very near each other, but not in contact. The plate to be magnetized is laid flat upon the magnets, so as to extend equally over the surface of both. The bars are then drawn asunder, till the plate just rests with its extremities in contact with the extreme poles of the two magnets, and then it is slid off sideways, and removed to some distance, preserving the parallelism of its position with that of the magnets till these are restored to the proximity with which the operation commenced. The process is repeated with the other side of the plate in contact with the magnets; and in the case of thin small plates,—such as I have adopted for the needles of sea-compasses—the condition of saturation is usually found to be obtained. Generally, however, to secure the maximum more satisfactorily, the plates are subjected to four strokes of the magnets, two on each side; and in hard short bars, six or eight strokes are usually given, partly on the edges, as well as on the flat sides. This process, for facility of reference, is distinguished as the method KS.

The method of magnetizing steel plates or bars now described, admits of such celerity in performing the manipulations, that a dozen plates, bars, or needles, adapted for compasses, may be easily magnetized to saturation in four or five minutes; and plates of sixteen inches to two feet in length, of various thicknesses, up to 0.1 or even 0.25 inch, may be thoroughly magnetized within a minute, — the thinner plates indeed, in half that time, or even less. In the case of such large plates, when a considerable number is required to be magnetized, I employ a long board with a straight channel formed on the surface by means of two laths nailed upon it. This gives additional facilities for effecting the manipulations by the guidance afforded to the magnets as they are separated; and by placing a short pin of brass wire in the middle of the groove, the poles of the two magnets are prevented coming into contact, whilst two other pins, at the extremity of range of the magnets, adjusted to the length of the plates under operation, serve to check the separation of the magnets at the exact distance required, and greatly to economize the time of the magnetizer.

So convenient and effective is this process, that by means of a pair of strong two-feet magnets only, compass needles or dipping needles of the usual form can be brought up to their maximum power *without removing their agate caps or centres;* and such bars can be thoroughly magnetized when covered with a thick coat of varnish. Compound needles, moreover, can, in this wise be strengthened, if accidentally reduced, without being taken asunder ; and the power of such needles can be increased after being originally put together; whilst ordinary compass needles, or those of Kater's compasses, can be well and quickly magnetized without detaching them from their cards or appendages. The process with the use of the pair of bars above described, is equally effective with massive bars of short lengths, *however hard in temper :* such as rectangular prisms of the hardest cast steel, 0·5 inch square, and 6 inches in length : or thick tempered bars of much greater masses. And it is likewise effective in developing the maximum power in small compound magnets, without the separation of the bars or plates.

A *discernible* advantage is yielded to this process, in common with what has been often observed, by raising the proximate ends of the magnets, so that the plates laid upon them may lie, not in general contact, but in a plane forming an acute angle with that of each of the magnets. But the gain, as observed at various angles from 2° 10′ to 8° 20′, was always found so small—never exceeding one-fiftieth part of the amount of power, and, on an average of the maxima of many trials with a variety of bars, being little more than one-eightieth,—that this modification, which is attended with some inconvenience in practice, and which gives no additional power in the case of combination, was not thought of importance enough to be adopted in preference to the method above described.

CHAPTER III.

AMONG the various methods in practice for determining the directive powers of magnetic bars or needles, that of Coulomb, by the balance of torsion, is generally considered as the most exact and satisfactory. But for facility in obtaining results, and for convenience in comparing the experiments of different persons with different instruments, the method of *deviations*—that is, the observing of the relative influences of the several bars on the needle of a compass, when acting at right angles to the natural position of the needle, and at a given or proportional distance—seems to me to possess great and peculiar advantages.

In this case, the bar under examination being laid in a horizontal position, and at right angles to the magnetic meridian, will be precisely in the plane of the magnetic equator of the earth;

c

so as to receive no inductive influence whatever from the terrestrial magnetism, and thus to exhibit only its own actual energy. And being placed at right angles to the suspended needle, the angle of deviation in the needle from the magnetic meridian will afford a very exact means of determining the directive energy of the bar, and a ready mode of comparison with other similar bars tried in the same manner; or even, though less accurately, with bars of different lengths if examined at proportional distances from the compass.

These deviations, in the case of experiments with magnets placed at considerable distances from the compass—say, at the distance of five or six times the length of the needle of the compass employed—will afford, in the proportions of their tangents, a proximate measure of the powers of different similar bars, of the same length, tried in the same manner, sufficiently near, even without correction, for most of the practical purposes contemplated in these investigations.

Could the compass needle, indeed, be made indefinitely short, or could the powers of the bars be determined at distances indefinitely great; then the tangent of deviations would afford exact measures of the directive powers of the several bars respectively.

For let c (*fig.* 1.) be a needle indefinitely short, and $c\,e$ the magnetic meridian, or direction of terrestrial magnetism. On placing a magnetic bar, m, at right angles to the natural position of the needle, at any given distance, its action will cause the needle to deviate in the direction, say of $c\,b$. Let the extent $c\,a$ represent the power, acting horizontally, of the terrestrial magnetism, which, in the same place, or under an equal magnetic dip, may be considered as always the same. Then, completing the parallelogram, $c\,a\,b\,d$, $c\,d$ will represent the directive power of the magnet m, at that particular distance. And this, with relation to the horizontally acting power of the earth's magnetism, will be as the tangent of the angle $a\,c\,b : 1.$*

For in the triangle $a\,c\,b$, in which we have $a\,b = c\,d$, $a\,c$ represents the directive force on the horizontal needle of terrestrial magnetism, and $a\,b$ the directive power of the magnet m; but $c\,a : a\,b : :$ radius : tangent of angle $a\,c\,b$. So that calling the directive force of the earth 1, the proportional force of the magnet will be represented simply by the tangent of the angle of deviation of the needle. Hence, if the directive

---

* In this investigation, the action of only one pole of the magnet is considered; because the results, as to proportions, are the same.

power of the magnet m, at the distance referred to, be to that of the earth as tangent of $a\,c\,b$: 1; the proportional power of any other similar magnet, m', of a different directive force, but of the same length, will be to that of the earth as the tangent of the deviation which, at the same distance from the compass, it produces; consequently, the different magnets will be to one another, in respect to directive power, as the tangents of the deviations severally produced by them.

Strictly, however, this law of the deviations cannot be realized in practice; as no needle indefinitely small can be made use of, nor can deviations at a distance, indefinitely great, be observed. Hence, the results obtained by this method, require, for strict accuracy, certain corrections; which at short distances may be very considerable in quantity, but diminish rapidly as the distances are enlarged. First, because the force m, is not truly exerted in the direction $c\,d$, but obliquely on either side, in two directions, from the foci of attraction near the extremities of the needle. And secondly, because the magnet does not act with the same power upon the two poles of the needle, in its deviated position, as it would were the needle restrained to a rectangular position; for the force exerted by the magnet on the needle, whilst sustained in a rectangular position, would be (considering only the effect of the nearest pole, which is sufficient for illustration) attractively on one pole of the needle, and repulsively on the other, the same; or just double the action of either. But the effect of the attractive and repulsive actions of the magnet on the same needle, when in a position of deviation, is different; the sum of the reciprocals of the two squares of the different distances of the poles of the deviated needle not being the same as double the square of the medium distance.

Nevertheless, the tangents of deviations, if the experiments be made at a proper distance, will be found, as from actual trials I have repeatedly proved, to afford results sufficiently near for practical purposes generally. So that for the determination of the relative energies of similar bars or plates—the changes in any magnets by time or circumstance—the superiority of one magnet or needle over another of the same, or nearly the same length—the relative advantages of the different modes of magnetizing, and other objects, the simple method of deviations is found to be most ready and effective. And even for bars not equal as to length, their deflecting energy being ascertained at distances proportionate to their lengths, will generally give results of useful approximation.

As, however, the equations will be very considerable, when deviations are taken at short distances, such as two or three times the length of the needle of the compass ; and as, for particular purposes, the exact directive powers proportionally of different magnets may be required, it will be useful to investigate the *nature* of the approximation to the real powers afforded by the foregoing method, and the value of the equations for obtaining true results.

This may be accomplished by comparing the observed results of the foregoing method, with those obtained by other modes which afford accurate proportions as to directive power; such as the method of oscillations with a bar of great capacity, examined when magnetized in different degrees of energy; or the method of torsion applied to the same bar in its different magnetic states. For such bar, in such several states, being examined by these different methods, the corrections of the method of deviations, due to each angle of deviation, will be given. But the same may be accomplished proximately, and almost precisely, by increasing the distance of the magnet under examination, till the proportional tangents of certain strong and weak bars, of the same length, become uniform when more remote distances are tried.

The method of deviations, however, can be

shewn to be capable of being so modified as to
afford, with bars of the same length, true
results,—*i. e.* by making the line of action per-
pendicular to the resulting position of the compass needle.   For in such an arrangement, the
equation in the former method, arising out of
the varying *distances* betwixt the foci of attraction in the bar and compass needle, under
different angles of deviation, altogether disappears; whilst the varying *direction* of the
forces acting on the needle, under different
deviations, in that method, becomes, for magnets
of the same length, and placed at the same
distance, an uniform direction, and so easily
capable of determination.

But the principle on which the exact proportion of the powers of magnets of different
intensities, but of the same length, are, by this
modified method of deviations, determinable,
may be more particularly stated :—

If a pendulum, or a magnetic needle suspended
on a pivot, be drawn aside from the position in
which, under the simple action of terrestrial influence, it is at rest, the force with which it is
solicited to return to its natural position will be
proportional to the sine of the angle of deviation.
Were it required, therefore, to sustain the pendulum or needle in certain given positions of deflec-
tion, and this were done by the adjustment of

equivalent statical forces acting always in the same direction in respect to the *deviated* position of the pendulum or needle, (say at right angles to its final position of deviation) then such statical forces, for different quantities of deviation, must be proportional to the sines of the angles of deviation respectively.

Let *s b*, (*fig.* 2.) be the needle of a compass, in this proposed condition, in which the action of terrestrial magnetism, in the direction *e′ e*, is controlled by the force of a magnet, *m*, acting in a direction at right angles to the deviated position of the needle; and, considering only the resultant action of all the forces of the magnet, as if comprised in the simple influence of one pole, and that acting on the centre of attraction of one pole of the needle,—let *a c* represent the force of terrestrial magnetism on the horizontal needle, and *c d*, or *a b*, the sustaining force of the magnet; then it is obvious that *c a* : *a b* : : sin.∠ *a b c* : sin.∠ *a c b*. Call the horizontal force of terrestrial magnetism E, and the resultant energy of the magnet M,—the angle of deviation D, and the angle opposite *c a*, F : then—

$$M = \frac{E. \sin D}{\sin F}.$$

If, however, no change be made in the dis-

tance or direction of action of the sustaining
force M, in regard to its rectangular position
with the deviated needle, then, in comparing
the power of M, with that of any other stronger
or weaker magnets, M', M'', etc., the sin. F.
will become a constant quantity; so as, if considered as a right angle, or equal to 1, not to
affect the *proportional* results as to the powers of
the different magnets; and, therefore, the exact
relation to the power of terrestrial magnetism
on the horizontal needle not being regarded,
but only the relation of the powers of different
similar magnets to one another,—which is the
thing chiefly essential,—this will be expressed
proportionally by the sines of the angles of deviation, respectively, when the compass needle is
sustained in the prescribed position.

Carrying out, therefore, this arrangement experimentally—which, with the simple apparatus
about to be described, is very easily accomplished—we have an easy and practical mode,
without the use of additional instruments, of
ascertaining the proportional directive energies
of all magnets, being equal in length, however
differing in mass or power; — a mode free from
the equations belonging to the other method of
deviations by rectangular forces, and calculated
not only to give the strict proportional powers,
but, by the application of the preceding formula,

to shew the amount of equations due to the other
method, and to enable us to form tables of equations for any particular distance, length of magnet, or size of compass,—for correcting, without
difficulty, the observed results of the method
proposed.

In this manner I have found that the *correction*
of the deviation, observed according to the method first recommended and generally pursued in
these investigations, was, on a powerful six-inch
bar magnet, examined at only twelve inches distance from a Kater's compass, with a needle of
4·9 inches, 1° 13′ in a deviation of 28° 26′; but
that at the distance of twenty inches — a distance abundantly within the limits of practical
observation—the correction almost entirely disappeared, when the deviation was 12° 30′.

But as this subject is of much importance
where accurate comparisons of magnets of very
different degrees of power are required, as well
as for giving confidence in inferences drawn
from determinations of the power of magnets
by the method of deviations,—it may be well,
in this place, to give an example of an experiment, undertaken for the twofold purpose of
determining the equations of the tangents due
to several different angles of deviation, and of
verifying, by a diversity of processes, the results
obtained.

For this experiment I employed four bars of steel, of very similar form and dimensions, but of different capacities for the magnetic condition, each being six inches long, and, in form, a rectangular prism, of 0·5-inch square. These after having been magnetized fully, were left for two or three days without protection, so that no change of energy was to be expected during the time occupied in the experiment. The power of each bar was now ascertained on the *trial-board*, in succession; first, by the method of tangential deviations, at the distance of twelve inches from the centre of the compass, and then by the other method of deviations, in which the needle is sustained in a position at right angles to that of the magnet. The differences of the results given by the two methods (the latter being corrected by the formula), were considered as the equations of the tangential method at the several deviations.

Another set of corresponding experiments, by both methods, was likewise made with each bar whilst in the same magnetic condition, at the distance of 19·25 inches from the compass; in which case the equation of tangents, it was certain, must be greatly lessened, and so a proximate series of corrections afforded for equating the results at the shorter distance.

If, now, the proportionate powers of the several bars, as ascertained at these two distances—
where the quantity of error, in respect to the
simple proportions of the tangents of deviation,
must greatly differ—should be found to be reasonably similar, this alone would have afforded
a fair verification, both of the principle adopted,
and of the results obtained by it ; but, for further assurance as to the accuracy of the proportions thus yielded, I tried the first bar (the
strongest in energy) by *the method of oscillations*,
when suspended by a few fibres of silk; and,
after having greatly reduced its power, and in
such a way that the distribution of magnetism
was very equable throughout, the rate of the
oscillations was again tried, commencing, in
both cases, at similar angles of deviation from
the magnetic meridian.   This bar being in the
latter condition, as to energy, also tried by the
two methods of deviations, and at the two particular distances adopted in the other instances,—
afforded the means of connecting the experiments of the extremes of deviations, with the
proportions determined by the oscillations.

The following are the results, as to the powers of the various bars, given by the different methods :—

I. Distance of Bars from the Compass, 12 Inches.						
Deno-mination of the Bars.	Tangential Method.			Equational Method.		
	Mean Devia-tion.	Tangent.	Reduced Propor-tions.	Mean Devia-tion.	Corrected sine : $\overline{E \sin. D}$ / $\overline{\sin. F}$ *	Reduced Propor-tions.
H 1	° 39·36	827	——	° ′ 48·18	756·5	100·0
T	28·26	541	——	30·29	514·1	67·9
E	23·25	433	——	24·25	418·7	55·4
S	17·48	321	——	18·20	318·7	42·1
H 2	12·10	216	——	12·20	216·4	28·6
II. Distance of Bars from the Compass, 19·25 Inches.						
H 1	12·30	221·7	100·0	12·46	222·3	100·0
T	8·38	151·8	68·5	8·43	152·4	68·6
E	7·2	123·4	55·5	7·6	124·3	55·9
S	5·16	92·2	41·6	5·18	92·9	41·8
H 2	3·36	62·9	28·4	3·33	62·3	28·0

Bar. H 1: performed 21 oscillations in 5 0′ ″=4·2 per min.
—— H 2: performed 135 (in two sets) in 17 4=7·91 per min.

* The quantity F, (angle a b c) at the nearest distance, was found by calculation to be 99° 17′, as derived from these data: viz. Distance betwixt the focus of attraction of the compass needle, and that of the magnetic bar, 12·7 inches;— this focal position, to which all the magnetic forces in the bar or needle may be referred, being found by former experiments to be about one-twelfth of the length of the magnet from its extremity. Focal length of the compass needle, 4·1 inches. The focal length of each arm from the pivot, 2·05. In the more remote position in which the distance betwixt the foci of the needle and bar, measured 19·85 inches, F becomes=95° 56′.

Hence, as the directive forces of this bar, in its different states, will be proportional to the squares of the number of oscillations in a given time, we have $7 \cdot 91^2 : 4 \cdot 2^2 : : 100 : 28 \cdot 19$, differing from the mean of the proportional powers of this bar in its two conditions, as derived from three sets of deviations observed at different distances, or by a different method, only $0 \cdot 14$, or about $\frac{1}{200}$th part of the whole. This accordance of results obtained by methods so different, was most satisfactory; whilst the agreement of the three sets of reduced proportions is equally striking.

Thus was fairly established, I think, the proposition primarily asserted;—that, for practical purposes generally, the tangents of the deviations produced by magnets of similar lengths, but of different energies, on a compass needle at a given sufficient distance (say, for particular comparison, not less than five or six times the length of the needle of the compass employed in the experiment), afford a satisfactory estimate of their proportional powers. So satisfactory, indeed, are the proportions obtained in this instance, that, on comparing the "reduced powers" given by the "tangential method," with those yielded by the "equational method," at the greater distance, we find the differences quite inconsiderable.

For facilitating the trial of the relative powers of bars or compass needles, on the method
herein proposed,—and for enabling the experi-
menter to make satisfactory trials of comparison
at subsequent times, however great the interval
—the following plan was found to be convenient.

A straight flat board (*fig.* 3.) is made use of,
on which the compass, and the magnet to be
tried, are placed,—the board being about two
feet in length, as fitted for the trial of compass
needles—and, if designed for proving larger
magnets, being at least three times the length
of the longest bars. A black line, A B, is drawn
along the middle of the board, from end to end;
and near one extremity a circle or circles are
described—having their common centre in the
middle line—of such diameter as to correspond
nearly with the size of the compass or compasses
which it may be convenient to employ in the
experiment. The circle, marking the place of
the compass, is divided by a line or diameter *n s*,
at right angles to the middle line of the board.
Small wire pins, or merely marks drawn across
the middle line, are used to indicate the distances at which the magnets to be examined are
to be placed. For the examination of the powers, however, of any large number of similar
bars—or for the testing of plates or bars designed
for the needles of compasses--I employ a move-

able ruler, c, a little longer than the bars to be
tried, having small brass pins, three at one end,
and two at the other, adjusted to the dimensions
of the bars, so as conveniently to retain them in
the position of the ruler, whilst the middle pin,
at the end nearest the compass, serves for the
accurate adjustment, as to distance, of each of
the succession of bars. This ruler revolves on a
short brass pin, o, as a centre, which is adjusted
so that the magnet, when in its place, shall have
its nearest pole at the exact distance of two or
more of its lengths from the centre of the compass ; and so that, by turning the ruler on its
centre, the other pole of the magnet will be presented to the compass, at precisely the same
distance as the first.

The board being now placed pretty nearly
east and west, on any convenient table free from
iron, at a sufficient distance from all attracting
substances, and all proximate magnets being
carefully neutralized by one another, or else
placed at a distance, and in such a position that
their two poles may act equally and compensa-
tingly on the compass,—the compass is adjusted
within the circle, (I have recently placed two
pins in the circle, with corresponding holes in
the bottom of the compass, for facility and accuracy of adjustment), with its meridional line in
the direction of the diameter n s. Observing

which way the needle is deviated from the meridional line of the compass, the proximate end of the board is slid gently round in the same direction, till the needle of the compass exactly coincides with its meridional line. In this position the direction of the needle will correspond with the diameter *n s* on the board,—shewing that the board is now precisely in an east and west (magnetic) position. Hence, when any magnet is placed on the middle line, or on the ruler when adjusted evenly upon the middle line, its power, according to the tangential deviations produced on the compass, will be very quickly observed; and by forthwith turning the opposite pole towards the compass, a mean of deviations will be given, calculated both for the correction of errors of adjustment in the board or compass, as well as for any inequality in the powers of the two poles of the magnet.

By means of this simple apparatus, the powers of any large set of similar bars can be determined with singular facility—very easily in 30 seconds for each bar; whilst I have repeatedly found that in large sets of plates, of six, sixteen, and twenty-four inches in length, I was able to magnetize the plates, try their powers, by both poles, and to register the results, at the rate of a dozen plates in about eighteen minutes. And besides this facility of trial, there is great

advantage in accuracy when it is required to
compare the powers of the same bars at distant
intervals of time, as they can be brought with
the greatest precision into the same position and
circumstances by the use of the board.

For the trial of the powers of bars by the equational method—or that in which the proportions
of power are given by the sines of deviation—the
same apparatus is used. But instead of adjusting
the board by the compass in an east and west
position *before* the magnet is brought near; the
magnet is *previously* laid on the ruler, and then
the board is caused to slide round till the direction of the needle, under the combined action of
the magnet and terrestrial magnetism, coincides
with the meridional line of the compass. The
magnet being then removed to a distance, the
return of the needle to its natural position shews
the angle of deviation of the board at which the
compass needle had been sustained at right
angles to the position of the magnet. The experiment is repeated with the other pole of the
magnet towards the compass, and by the mean
of the results the inductive force of terrestrial
magnetism on the magnet (the magnet not being
in this case in the neutral or equatorial plane)
is exactly compensated.

CHAPTER IV.

THE DETERMINATION OF THE RELATIVE STRENGTH, OR TE-
NACIOUSNESS OF THE MAGNETIC CONDITION, IN EACH OF
A SERIES OF APPARENTLY SIMILAR PLATES OR BARS OF
STEEL.

———

TENACIOUSNESS or fixidity, which is a grand element in good magnetic instruments—especially in sea-compasses—has heretofore been comparatively little attended to in their construction; the great point generally aimed at having been, to obtain, in the first instance, high magnetic energy. And thus the needles of *compasses*, in which *permanency* of power is of very high consideration, have generally been constructed on a principle incompatible with the attainment of this most important property. For not only may needles or bars on the ordinary construction and temper, exhibit, in many cases, an effective energy at the first; but even a perfectly soft or untempered bar of good steel,

of a certain limited mass, will have a very considerable capacity for magnetism, and so much
apparent strength as to yield an efficient power
for a compass needle, after being removed from
all extraneous aids for the retaining of the
power. But in such a needle, the magnetic
condition, as is well known, will be comparatively transient; and, indeed, so unfixed in its
character, that the mere proximity of another
magnet may be sufficient to neutralize, or even
reverse its polarity.

The method of determining the degree of
strength or tenaciousness, which I have now to
propose—and which will be found, I trust, at
once satisfactory, effective, and of most extensive applicability in the arts—resulted originally
from the subjecting of a large series of magnetized steel plates to a similar condition of violence, and then ascertaining, in the case of each
plate, the proportion of the remaining power.
This investigation, in its original form, was made
on several series of large steel plates, in which
the state of violence, constituting the test of
tenaciousness, was derived from the power of
the series itself, when formed into a fasciculus,
with the corresponding poles of all the plates
contiguous. My first experiment was this :—
Desirous of ascertaining what might be the effect, on the augmentation of the magnetic power,

of an extensive combination of steel plates; I
procured a large quantity of such plates (59 in
number), and had them reduced to common
dimensions, viz. 17·5 inches in length, and 1·33
in breadth; the thickness, which was exceedingly equable, being about 0·03 inch. All
these, I magnetized to saturation, and determined their directive energy separately. This
was such as to produce a mean deviation of
about 23° on the needle of a compass to which
each plate was presented at 17·5 inches, or one
length, distance from its centre, and in the line
of east and west of the compass.

The whole of the bars were subsequently
placed upon each other, with their corresponding poles in contact, so as to form one powerful
compound magnet. After allowing them to
remain in this position of violence for about a
couple of hours, (I ultimately found that as
many minutes, probably seconds, would have
answered just as well,) the combination was
dissolved; and each of the plates subjected to
experiment for the determination of its resulting
energy. It was then found that some of the
plates had their polarity reversed—in some, it
was totally neutralized—in many very feeble;
whilst only twelve of the whole number retained
a deflecting energy of from 14° to 18°.

But these plates were of ordinary manufac-

ture. I next tried the effect of combination in a
large uniform series of very beautiful plates of
cast steel, tempered throughout their length,
of two feet long, 1·5 inches broad, and ·042
inch thick, and weighing, on an average, 2869
grains.

Thirty-two of these plates being magnetized
to saturation by the process K.S, exhibited a
mean power of deviation on the compass, at one
length distance, as tried separately, of 16° 10′;
the weakest causing a deviation of about 15°,
and the strongest of about 18° 30′. All these
were then placed for a short interval in one
fasciculus, with their similar poles together,
and, before being entirely separated, the several
plates were alternately changed as to their position, and transferred to different parts of the
mass. The whole series being now separately
examined again, the average deviating power
was found to be reduced to 7° 25′.

The various plates, after this deterioration by
contact, exhibited the following differences of
condition.

6 plates retained a power of 10°+ to 11°.
8   ,,   ,,   8 + to 10.
8   ,,   ,,   6 + to 8.
4   ,,   ,,   4 + to 6.
1   ,,   ,,   2 + to 4.
5   ,,   ,,   0 + to 2.

On this process being variously repeated, it was found, that the several bars could thus be satisfactorily tested as to their respective permanency or tenacity of retention; for with rare, and probably accidental, exceptions, the results of different experiments, made with the same series and number of plates, were very generally accordant.

The following table, exhibiting the results of this test of tenaciousness, etc., on two sets of small plates designed for compound compass needles, as determined by the violence of their respective series, will serve both to illustrate these observations and to shew the efficacy of the test. Series A, it may be premised, consisted of eighteen very thin plates of tempered cast-steel, of 6 inches in length, ·425 inch in breadth, and weighing, on an average, 97·5 grains. Series B, consisted of nineteen plates, 6 inches in length, and 042 in breadth, and of 162 grains weight each.

Series A : Plates of 97·5 gr.				Series B. Plates of 162 gr.			
Nos. of the Plates.	Mean Deviation.		Series according to Strength.	Nos. of the Plates.	Mean Deviation.		Series according to Strength.
	When first Magnetized.	After Combination of 18.			When first Magnetized.	After Combination of 19.	
I.	6·20	3·0	11	I.	10·0	1·45	19
II.	6·22	3·46	2	II.	12·20	5·48	3
III.	6·30	3·8	8	III.	11.30	3·56	11
IV.	6·50	2·40	15	IV.	12·50	5·10	5
V.	6·33	3·20	5	V.	10·20	2·15	18
VI.	6·30	3·18	7	VI.	12·0	4·0	10
VII.	7·55	1·30	18	VII.	12·20	3·5	15
VIII.	6·33	3·20	6	VIII.	13·0	5·45	4
IX.	6·20	3·0	12	IX.	12·40	5·48	2
X.	5·45	3·30	4	X.	12·42	3·40	12
XI.	6·50	3·0	10	XI.	10·15	2·35	16
XII.	6·18	2·35	16	XII.	12·15	4·5	9
XIII.	6·25	3·7	9	XIII.	12·0	2·15	17
XIV.	6·20	2·52	13	XIV.	12·10	4·20	8 } 6 }
XV.	6·6	2·27	17	XV.	11·40	4·30	7
XVI.	6·10	3·46	3	XVI.	10·30	4·35	6 } 8 }
XVII.	6·45	4·14	1	XVII.	10·15	3·30	14
XVIII	6·30	2·45	14	XVIII	11·0	3·30	13
				XIX.	12·30	7·50	1

For the verification of the results thus yielded, as to the *strength* or tenaciousness of the plates, I arranged series A, in sets of six plates each, according to the order of the powers, as determined by the test, commencing with the strongest and proceeding in regular progress to the weakest: the plates of each set being re-magnetized, and then combined, the following satisfactory results, as to the proportional powers of the different sets, were obtained.

No. of the set in the order of tenaciousness.	Weight of the six Plates. Grains.	Deviation at two lengths.
		o '
I.	566	22·25
II.	590	21·28
III.	595	21·14

In like manner, the series B was tried in combinations of four plates, and the accuracy of the results, as to strength, similarly verified.

No. of the set.	Total weight. Grains.	Deviation at two lengths. o '
I.	644	25·0
II.	652	23·35
III.	643	22·40
IV.	—	21·4

Hence were the results, as to the individual tenaciousness of the plates, verified by the consistency of the powers in combination in these smaller series; the most tenacious plates yielding in similar combinations the most considerable powers, and the least tenacious the lowest powers.

For the testing of plates of this kind, or the bars or needles of compasses of uniform, or nearly uniform, dimensions, I have found it both convenient and advantageous, for comparison, to make use of a powerful highly tempered—or what is still better, a *perfectly hard*—bar magnet, of length and width corresponding pretty nearly with these dimensions of the plates to be tried, for a TEST-MAGNET.

The test-magnet which I usually employ for these small bars, etc., consists of a rectangular prism, of best cast steel, thoroughly hardened throughout the mass, six inches in length and 0·5 inch square. Its power is great, occasioning a deviation of 38° to 39°, on a compass, at twelve inches or two lengths' distance. But when employed for testing, I first *reduce* its power, by laying upon it a similar bar with corresponding poles coincident, which brings its deviating power down to about 33° or 34°. In this reduced state, the testing of any number of compass needles, or other small bars, produces no further deterioration; so that the degree of violence to which each is subjected, may be considered (except as to slight differences in the nature of the contact) as *precisely* similar.

The process of testing compass needles, plates, or bars — which, by a due adaptation of the test-bar, may also be applied to cutting instruments of various kinds—simply consists in laying each of the several bars, etc., after being thoroughly magnetized, in succession upon the test-bar, with similar poles in contact. The mere momentary contact of the two sides of the plate, or the several sides of a thick bar, is sufficient,—*care being taken to bring the plate or bar evenly down upon the test-magnet, without sliding or friction*, which would augment the tendency to neutralize or change the polarity.

The time required for the whole routine of
this process is only about a minute, to a minute
and a half, for each plate or bar : and the precision of the results is such that the whole
series, though amounting to several dozens, can
be satisfactorily arranged, in the order of their
relative tenaciousness or strength, in a numerical succession !

By comparing the results thus obtained with
the powers of the bars or plates, respectively,
when first magnetized, the very best can be
selected, as to their adaptation for compasses, or
other particular object ; those being considered
as the *best*, in which the product of the forces
of the original power and the reduced power is
the greatest.

The importance of this test is further shewn
by the surprising differences which it detects
in, apparently, most similar bars ; for, in sets
of bars or plates—constructed out of the same
mass of steel, wrought by the same hand, and
tempered, as may be supposed, precisely alike
—the greatest differences will be found ;—such
that, whilst some may sustain the loss of not
more, perhaps, than one-sixth of their maximum power, others will lose the whole of their
magnetism, or even have their poles reversed.
Yet the manufacturer, himself probably, was not
at all aware of the difference ; nor could he, by

any decided or satisfactory means, separate the good from the bad. Hence in the procuring, from time to time, of the apparatus for these investigations—though a first-rate manufacturer was employed, and the best talent and inspection of his extensive establishment engaged for rendering the various bars and plates as perfect as possible,—very many of the articles, individually, as well as some whole sets, proved not only defective, but *utterly useless* for the object contemplated. Yet with all these deficiencies —necessarily pertaining to the difficult process of tempering, and the varying qualities of steel —it is but justice to the manufacturer to say, that I had not before ever been able to procure plates and bars of such difficult construction, so beautiful, exact, and good.

One example, in conclusion of this chapter, may be usefully adduced, as shewing the strikingly discriminating efficacy of this method of testing, in regard to compass needles. From a number of needles, furnished me from the Navy stores, for trial and comparison, I 'select two for illustration,—(No. IV.) one of the best out of nine, and (No. VII.) one of the worst. They are both by the same compass-maker, and precisely of the same form and appearance. These I compare with two bars of my own,—in form, plain parallelopipedons, and very nearly resem-

bling, in all their dimensions (but without any
hole in the centre), the compass needles described.

The four bars having been magnetized by the
process K. S., their maximum powers were severally determined by the deviation produced on
the compass at two lengths' distance, and then,
having subjected them all to the test, their
resulting energies, for ascertaining their relative tenaciousness, were determined. The following are the particulars of the experiment :—

Description of Bar or Needle.	Weight in Grains.	Maximum Power.	Power after the Test.
		o ′	o ′
Navy Needle IV.	510	13·30	— 1·0
Navy Needle VII.	580	5·35	— 3·30
Bar . . . I.	560	19·30	+ 16·40
Bar . . . II.	656	22·20	+ 17·0

But here the test-bar is found *too powerful* for
a fair comparison of the relative degree of
tenaciousness of the different bars,—the poles of
both the Navy needles being changed. Hence
a test-bar of weaker power becomes, in this
case, desirable; or, instead of changing the
test-bar, its action may be conveniently diminished by the interposition of some solid substance, betwixt the test-magnet and the bars to
be tested, to prevent contact. Generally, an
exceedingly thin plate of brass, or slip of ivory,
will be found convenient; but the action of the

test, in the example adduced, is shewn more
strikingly by employing a thicker substance ;—
I therefore give the result of the testing through
a piece of plate glass, 0·18 inch in thickness.

Description of the Needle or Bar.	Maximum Power.*	Power as reduced by Test,—0·18 inch separated.
Navy Needle IV.	13·30°	4·5°
Navy Needle VII.	5·35	— 1·48
Bar . . . I.	19·30	+ 17·40
Bar . . . II.	22·20	18·45

These results exhibit, by the way, the exceeding defectiveness of quality in the magnets
hitherto generally used in sea compasses ;—
seeing that, in one of the best out of nine,
taken as fair specimens of those in the Navy
stores, the capacity for magnetism is only about
two-thirds as great as that of the plain bar,
(No. I.), and its power of retention only about
one-seventh !

---

* The maxima powers were here assumed from the former
trials,—these being known to exhibit, very nearly, the relative
energies of the bars when thoroughly magnetized.

CHAPTER V.

THE DETERMINATION OF THE RATIO OF THE POWER OF MAG-
NETS, WITH THE AUGMENTATION OF THE THICKNESS OR
MASS, IN OTHERWISE SIMILAR BARS; WITH THE MODIFI-
CATIONS OF SUCH RATIO BY DIFFERENCE OF TEMPERING.

ONE of the inferences of Captain Kater, deduced
from his investigations on "the best kind of
steel and form for a compass needle," was, that
the directive force "in needles of nearly the
same length and form, is as the mass." Other
magneticians have come to a different conclusion. And, in the recent personal investigations already referred to, (Chap. I.) the results
were quite decisive as to this—that the magnetic
powers of bars, similar in all respects, excepting
as to thickness, *are not proportional* to their respective masses; but that the ratio of augmentation
of power diminishes as the thickness increases.

But the series of bars employed in these early investigations, being unequal in quality and temper, was not sufficient for the determination of the ratio of power with the mass. This subject, however, being of much importance in practical magnetics—and especially in the construction of compass needles of single bars, where the power of the needle, and consequently its weight, should have a due relation to the weight of the card and appendages to be directed—I instituted a series of experiments for the determination of the ratio of the power in respect to the mass, in bars of the same quality and dimensions, except as to thickness, combining, with the inquiry, the consideration of the effect of difference of tempering.

As the kind of tempering for magnets and compass needles, prevalently adopted by the instrument makers and manufacturers in this country, is that of a moderate hardening only at the ends, whilst all my own investigations had tended to shew a decided advantage in tempering from end to end of the mass,—I proceeded, in the first instance, to determine the respective ratios of power in bars of different thicknesses, in graduated series, tempered in both ways. For this experiment I procured a double set of bars of best cast steel, *tempered throughout* at a blue, or spring temper,—each bar being exactly six

inches long, and half an inch broad,—and com-
prising a series of $\frac{1}{2}$, $\frac{1}{4}$, $\frac{1}{8}$, $\frac{1}{16}$, and $\frac{1}{32}$ of an inch
in thickness.   A second series, as nearly as
possible similar in dimensions and quality, but
tempered only to the extent of about an inch at
the ends (the intermediate portion of the bars
being soft), was likewise procured at the same
time.

The bars of each series were then magnetized
to saturation, with a pair of powerful two-feet
bar-magnets, by the process K S, several different times, and their powers by the method of
deviations, on each occasion (in seven or eight
trials) determined, at two lengths', or twelve
inches distance, from the centre of the compass.
The extreme differences observed in the several
trials of the same bar, seldom exceeded one-
twentieth of the whole power.

The powers of these several bars projected
in ordinates, according to the tangents of the
deviations which they respectively produced,
with their weights for abscisses — as hereafter
described—afforded two series of points, by the
guidance of which two curves were drawn,
representing, proximately, the various powers of
bars of every measure of thickness, in the different kinds of tempering, up to the maximum
mass of a rectangular prism of half an inch
square.   These curves exhibited a very decided

E

advantage, as to power, in the tempered series
—except in the very thinnest bars—and, generally, there was a satisfactory accordance with
the relative ordinates; but not in all cases.
Hence, though the bars were constructed with
special care, by an able and intelligent manufacturer, yet, from these observable discrepancies, from the proximate curve, it was inferred,
that the temper of some of the bars was very
defective.

In the case of bar No. III., the third in the
fully tempered series, the thickest being marked
No I. (see the following table), the discrepancy
was specially remarkable—the termination of
the ordinate given by its magnetic power falling
below the curve about one-eighth of the entire
ordinate.   As the curve representing the powers
of the harder series of bars, was, in this
place, considerably above that of the other
series, it was inferred (from the view of the
nature of the magnetic condition in magnets
described in a subsequent chapter) that the
temper of this bar must be considerably too
low.   And on trying the strength, or power of
retention, of the various bars—by subjecting
the whole of the series to a test of the same degree of violence—it was found to be exactly as
anticipated; for whilst the bars Nos. I. and II.,
lost, severally, betwixt 30 and 40 per cent. of

their original power by the test—both of the bars, No. III., lost above 90! Hence the exception served to establish the rule—as to the efficiency of the method of testing. In fact, these bars, as well as IV. and V. of the same series, which were now found to be weaker in tenaciousness, proved to be actually softer, in temper, than the corresponding bars of the series tempered only at the ends.

Hence, from this accordance of the degree of tenaciousness with the hardness of temper, was suggested, *the method of determining the actual degree of temper* in bars or instruments of steel,— more particularly described in the next chapter— a method, it is believed, which will enable us to discriminate such small differences as no other known test, nor indeed any measure of experience in cutlery, or other manufacture of steel, could be sufficient for discovering.

The following table,—exhibiting the results obtained with the different bars above described, in their *highest* magnetic condition, as to energy, and also as to their tenaciousness—is now introduced, (though some of the particulars will require to be repeated) both for the verification of the foregoing remarks, and for exemplifying the process pursued for obtaining improved data for the projection of curves of temper or hardness.

No. of the Bars.	Weight in Grains.	Mean Deviation at Two Lengths.	Reduced power after the 15 bars being laid together.	Improved state after being laid in pairs by contrary poles.
I.	II.	III	IV.	V.

Bars tempered throughout.   Series T.*

		°	°	°
I 1	2860	29·10	21·20	24·0
2	3018	31·20	23·50	26·28
II 1	1535	27·7	22·50	23·48
2	1530	28·45	23·52	25·0
III 1	666	18·27	3·5	3·50
2	635	18·15	—0·7	+ 0·46
IV 1	312	13·30	+ 2·48	3·46
2	317	14·18	5·35	5·56
V 1	158	11·0	2·40	2·57
2	160	10·52	1·0	1·24

Bars tempered at the Ends.   Series E.

I	3030	24·50	11·35	16·10
II	1518	19·48	4·38	6·5
III	670	18·30	4·32	6·15
IV	314	15·0	5·50	6·52
V	160	11·13	4·50	5·12

In order, now, to determine still more satisfactorily the ratio of the power with the mass, in the bars of the particular kinds of tempering thus subjected to experiment—as well as to

* Col. IV. shews the state of the bars after being tested by their own mass with similar poles coincident, and variously shifted in their respective places in the pile.   Col. V. is introduced to shew the *elasticity* of the magnetic condition, and its tendency to recover a part of its loss:—indicating that whereas in the maximum state (Col. III.) the tendency is to lose power,—here, in the reduced state by the test, there is a contrary tendency.

verify the inferences which had been made as to
the discrepancies observed being due to defects
of tempering in certain bars,—I procured new
sets, very carefully tempered, of Nos. III. and
IV. of the harder series T, and of No. II. of
the softer series E. The results obtained with
these new bars satisfactorily proved the correctness of the original conjecture as to the cause of
the discrepancies, as clearly appears on comparing their powers with those of the defective
ones which they were designed to replace. The
following are the several relations, necessary for
comparison, in the defective and improved bars.

Original Series.				Improved Series.			
Denomi-nation of Bars.	Weight in Grains.	Maxi-mum Power.	Reduced power by Test Bar T. I.*	Denomi-nation of Bars.	Weight in Grains.	Maxi-mum Power.	Reduced Power by Test Bar T. I.
T. III. mean of 1 & 2.	650	18·21′	9·25′	T III. 3	710	22·0′	16·0′
T. IV. mean of 1 & 2.	314	13·54	8·4	T. IV. 3	318	17·5	12·45
E. II. 1.	1518	19·48	9·25	E. II. 2	1467	27·15	22·20

These results were sufficiently available for a
satisfactory improvement of the curves of powers

* The Test-Bar, T. I, had a power during the operation of
testing of 25° 20′ deviation, at two lengths' distance.

belonging to bars tempered throughout, at a spring-temper, and to a similar series partially tempered in the same way—only at the ends.

But being desirous of carrying the investigations, as to the curves afforded by different degrees of tempering, to still more conclusive results, and with the full expectation of being able to obtain powers with very hard bars, far beyond what such a condition in steel has hitherto (I believe) been considered by magneticians to be capable ; I procured two other sets of bars, corresponding in dimensions and quality, as also in the nature of the series, with those already described. One of these consisted of a series of five *perfectly soft* bars—softened in the fire after being forged—and the other of a series hardened to the utmost degree.

These, on trial, gave results, respectively, of so regular a ratio, that—with one principal exception in the case of the bar No. IV., of the hard series—they afforded proximate curves, exhibiting a very satisfactory accordance with the extent of the ordinates. And this exception, as in other instances, served only to verify the law; for on that bar being subjected, in common with the rest, to a severe test, it was found to lose about seven times as much power as any other of the series. In fact it was found to be quite soft—having probably been overheated, and so in measure decarbonized, in the attempt to harden it.

I was now in a condition to select out of the duplicate and triplicate bars, constructed of the same denomination, an improved set from which I might project more correct curves of the powers of the tempered series—as also curves to shew the comparative powers of the different series' of other degrees of hardness. In making the selection referred to, I was guided, not merely by the relative powers of the bars when magnetized to the utmost, but, mainly, by the accordance in respect to ratio, of the degree of tenaciousness of the bars of each series, as determined by the quantity of loss per centum, on subjecting the whole of the bars of each series to a similar test.

The magnet employed for this purpose was No. I. of the hardest series of bars (the same as is described in Chapter IV.), which is discriminated as *test-bar H*. And this was prepared for the purpose by reducing its power, as much as it could be constrained to give way, by subjecting it to the greatest measure of violence which it would have to bear: and that was by placing it in contact, of each of its sides in succession, with the hard bar No. II., when fully magnetized. The effect of this, in the instance referred to, was to reduce the power of the test-

bar from a deviation of 38° 20′, at two lengths
from the compass, to a deviation of 33° 15′; and
so firmly was this measure of power fixed in the
bar, that, on trial, after the testing of the whole
of the four series, amounting to twenty-seven
bars, the deviation produced was found to be
scarcely at all changed—being still 33° 10′.

All the bars having been once more carefully
magnetized to the utmost, by the process K S.
(for *no* advantage was gained by adopting the
process Æ S., even in the case of the thickest
and hardest bars) were now tried, as to their
ultimate powers or capacity ; and then, being
applied in succession to the test-bar, their
reduced powers were determined.

The following table exhibits these final
results.

No.	Weight of bars in grains.	Highest power.		Reduced power by test-bar H.		Tenaciousness.	
		Mean deviation.	Tangent.	Deviation.	Tangent.	Difference of Tangents.	Loss per cent.
I.	II.	III.	IV.	V.	VI.	VII.	VIII.

HARD SERIES, H.

		o ′					
I.	2950	38·20	791	—	—	—	—
II.	1404	31·50	621	28·15	537	84	13·5
III.	786	27·13	514	23·54	443	71	13·8
IV.	340	14·15	254	1·5	19	235	92·5
V.	175	10·0	176	9·8	161	15	8·5

TEMPERED SERIES, T.

I.	3018	31·50	621	22·30	414	207	33·3
II.	1535	27·10	513	18·0	325	188	36·6
III.	710	22·0	404	9·43	171	233	57·7
IV.	318	17·5	307	6·16	110	197	64·1
V.	160	10·48	191	2·3	36	155	81·1

TEMPERED AT THE ENDS, E.

I.	3030	24·30	456	10·22	183	273	59·9
II.	{1518 {1417	{19·50 {27·15	} 361 515	6.32 16·52	} 112 303	249 212	68·1 41·1
III.	670	18·13	329	3·32	59	270	82·2
IV.	314	14·40	262	3·6	54	208	79·5
V.	158	11·5	196	1·3	18	178	90·8

SOFT THROUGHOUT, S.

I.	2853	18·50	341	3·37	63	278	81·5
II.	1393	16·32	297	0·45	13	284	95·6
III.	713	15·8	270	—1·35	—28	298	110·4
IV.	317	11·30	203	—3·22	—59	262	129·1
V.	167	9·13	162	—2·50	—49	211	130·0

In the powers of these several classes, or series, of bars, we have a very observable, I may say satisfactory, regularity of ratio—whether we consider the capacity for magnetism, or the proportional loss of power, on application to the

test-bar, of each particular series, in relation
to its degree of hardness. Thus, omitting the
powers of the two thinnest bars of each series,
we find, in all the rest, an undeviating accession of power, or capacity, with the increase of
hardness ; and a still more remarkable accordance in tenaciousness or strength, increasing
from the softest to the hardest, with such striking
regularity, that we find but one deviation from
that order, in respect to bars of similar masses,
in the whole of the four series; and that is in
the instance of Bar IV., Series H., before alluded
to, which obviously belongs to another class as
to temper.

Another result, exhibiting a near approximation to regularity, is observable, in the action of
the test-bar in producing a gradually diminishing effect, as to proportional deterioration, as
the bars increase in thickness. [See the preceding table, col. VIII.] Thus, for example, in
series S., in which the equality, as to temper,
was known to be the most uniform—the whole
series having been *softened* in the fire after
being forged—the loss, per centum, is seen to
be the greatest in the thinnest bar, and to
become less and less, without one deviation,
up to the thickest of the class ; and the *tendency*
to a similar order is observable in all the other
classes, except the hardest, from the discrepan-

cies in which we are guided by analogy to assume, what the manufacturer is well aware of,
the exceeding difficulty of obtaining a full and
equable degree of *extreme* hardness, (especially
in thin plates) where over-heating and under-
heating are both detrimental to the required
result—the former, I imagine, by its decarbonizing influence—the latter, by its not affording sufficient mobility in the metal for the proper
arrangement of the particles.

This difference in the proportion of deterioration, indeed, was a circumstance to be expected,
as the test-bar must obviously have greater
power over bars of very inferior thickness and
energy, than over those of corresponding mass,
or approaching in equality of power, with itself.

The deviations from regularity in the measure
of deterioration, from the action of the test-bar,
with relation to the mass in the same series,
were not, however, to be considered as accidental; but were to be ascribed, as we have
intimated, to difference of tempering, that is,
where the steel of the bars might be considered
as similar. Hence, *before* projecting the results
in ˙proximate curves, I was led to consider the
principal deviations from regularity, in the proportional loss by the test-bar, from which it was
anticipated, that No. IV. H. [see the foregoing
table, col. VIII.] would be found below the curve

representative of its own series ; one of No. II.
E. considerably below, and the duplicate of
the same very much above, its own curve ; and
No. III., also of this series, below the mean
ratio. And *all* these anticipations—as will be
apparent on the inspection of the diagram, pl.
II., (in which, however, the mean of the powers
of the duplicate bars E. II., is only given)
were fully verified.

In this diagram the powers of the several
bars of the various series, according to the tangents of deviations (col. iv of the foregoing
table), are projected in ordinates, on a scale of
half an inch to tangent ·100, or five inches for
the tangent 1·000, being that of 45° ; and the
weights of the bars are given on a scale of a
quarter of an inch to 100 grains, as abscisses.
The proper line of abscisses of all the curves
is divided into a scale of grains weight ; whilst
four other scales are appended beneath, representing the actual weights of the bars of each
series. The terminations of the various ordinates
are marked by dots, and the class or series of
bars to which the dots severally belong, is distinguished by the discriminating letter of the
particular kind of tempering. Through, or
near, the points of termination of their respective
ordinates, the different curves are drawn, representing, proximately, the ratio of power with

the augmentation of the thickness, or mass,
in bars of a similar degree of hardness; and,
by comparison of the curves, are shewn, the
modifications of such ratio, as produced by four
different degrees of hardness.

I have here given the curves according to the
unequated deviations, because of the advantage
afforded by so doing in determining the *temper*
and *quality* of steel—hereafter to be exemplified
— by simple and direct reference, as well as
because of the facility which the diagram thus
affords for comparison betwixt the powers of any
other bars, or combinations of plates or bars,
with these employed in the present investigations.

In order, however, to the ready determination
of the *correct proportional powers* due to any
particular bars, or to the proximate curves, one
with another, where particular precision, as to
the proportional relations, may be desirable—I
have added to the diagram *a scale of equations*
derived from a series of comparative experiments by the modified method of deviations
described in Chap. III. p. 23, and verified by
the singularly accordant results obtained by
other experiments, as given at p. 29,—which
equations being set off *downward* from the
curves, in the direction of the ordinates, will
afford, for any particular tangent, or portion
of any curve, or projected power of any bar,

the *correct* proportional position of such tangent, curve, or projected power.

From the results now exhibited, in the projection of the powers of magnetic bars of different masses, and of different degrees of hardness, is distinctly shewn—may I not venture to say, to demonstration?—the error of the system most prevalently acted on in the construction of magnetical instruments, generally;—a system grounded on an erroneous supposition as to the capacity of steel of different degrees of hardness for the magnetic condition. As to the imagined superiority of a moderate hardening of the *ends only* of bars designed for magnets—the method adopted, I believe, by all our instrument makers, whether of sea compasses, or of artificial magnets, or other magnetical apparatus,—we here perceive the very small extent, as to mass, to which such a mode of tempering can, in any point of view, be beneficially applied. For it is very observable, that such a mode of tempering possesses no advantage as to capacity, though it has much disadvantage as to tenaciousness, except in very thin bars—not exceeding, on the scale of those under investigation, the fortieth of an inch in thickness, or about 130 grains in weight.*   But in larger masses,—in bars of

* The late talented author of the " Bakerian Lecture," read before the Royal Society, February 1, 1821—Captain Kater—

the same length and breadth,—we find such an advantage in the moderate hardening of the steel *throughout*, as to yield a superiority amounting, in the case of the heaviest bar of the series, to three-tenths more power than that of which a similar bar, as to mass and quality, tempered only at the ends, is capable.

As to the generally supposed *inferiority* in capacity for the magnetic condition of perfectly hard steel, we find a similar error. That thoroughly hard steel, indeed, is susceptible of less power than tempered or soft steel when first magnetized, *to a very limited extent of mass*, is, apparently, the fact; but as a general proposition it is most erroneous. For in all masses above the weight of 130 grains (that is, of the form and length here under consideration) *perfectly hard* steel appears to be superior in capacity to *soft* steel; in masses above 250 grains' weight, superior to bars tempered only at the ends; and above 400 grains, superior to any of the kinds

found that needles for compasses softened in the middle after being moderately hardened throughout, and left moderately hard at the ends only, were the most powerful. But the particular fact derived from Captain Kater's experiments, which he assumed as a general law—we now find, from the foregoing investigations, was only true in the case of thin or light needles such as he employed. And a similar mistake, we shall subsequently be able to prove, has been fallen into in respect to the best quality of steel for compass needles.

of tempering with which it has been compared. And so great is the superiority that, according to the diagram, it appears, that in the largest mass—being a square prism of half an inch— the power of the hard bar H 1, is to that of bars of equal weight (reckoning each bar 3000 grains) in the series T, E, and S, as 100 to 76·6, 58·3, and 40·9, respectively—or, if corrected for the equation of the tangents, as 100* to 79·0, 61·1, and 44·1 nearly.

Another erroneous idea, as prevalent, per-haps, as either of those mentioned above, is, that there is a peculiar, if not an insurmountable difficulty, in magnetizing, to any high degree of energy, very hard bars, if of considerable thickness; whereas I find no difficulty whatever in magnetizing, by the process K S, described in Chapter II., any of these yet subjected to the operation, even up to the largest mass of half an inch in thickness.

* The actual capacity of the hard bar is no doubt higher than the proportion here given to it—for on being repeatedly magnetized its power has been already augmented one-twentieth. The same treatment with other bars gave *no* increase. The curve of the hard bar, therefore, in the diagram, may require some correction, for its adjustment in due relation to those of the other series.'

CHAPTER VI.

THE DETERMINATION OF THE RELATIVE DEGREE OF HARD-
NESS OR *TEMPER* OF PLATES OR BARS COMPOSED OF THE
SAME KIND OF STEEL.

———

THE investigations of the preceding chapter
afford, in their results, the principle on which
the relative degree of hardness (on a mean of
the whole mass of each bar) of apparently
similar plates or bars, may be satisfactorily
determined. This is accomplished, in practice,
by observing the quantity of the deterioration of
power in any series of similar bars or plates,
either by the violence of combination in one
mass with similar poles coincident, or by the
simple action, on each bar or plate, of the test-
magnet. The bars or plates, so tested, which
lose the *largest* proportion of their maximum
power, are thus shewn to be the *softest*, and
those suffering less and less deterioration to be
progressively harder. Hence in the series of

F

tempered steel plates, B. (see table, at page 40) tested by the combination of the whole series, consisting of nineteen plates, in one fasciculus, we find the plate No. XIX. to be the hardest, and No. I. the softest of the whole series. And the arrangement of both the sets of plates, series A and B, as given numerically under the title of "Series according to strength," equally represents the order of the plates as to their relative condition of hardness. For the whole extent of the investigations on the qualities in steel on which the permanency of the magnetic condition is dependent, has tended to shew that the degree of hardness and that of tenaciousness are co-relative; this latter quality of the magnet being, apparently, the simple consequence of the hardness.

Ample verification of the accuracy of the principle on which this method of testing the *temper* of steel bars is founded, has already been afforded, I conceive, in the results of the preceding chapter. By arranging the table, however, given at page 57, in a different order (as annexed), the various interesting relations of temper and tenaciousness will appear more obvious and striking.

Denomination of Series.		Bars No. I. Weight, 2853 to 3030 grains. Tenaciousness			Bars No. II. Weight, 1393 to 1535 grains. Tenaciousness			Bars No. III. Weight, 670 to 786 grains. Tenaciousness			Bars No. IV. Weight, 314 to 340 grains. Tenaciousness			Bars No. V. Weight, 158 to 175 grains. Tenaciousness		
		Maximum Power.	Reduced by Test.	Loss per Cent.	Maximum Power.	Reduced by Test.	Loss per Cent.	Maximum Power.	Reduced by Test.	Loss per Cent.	Maximum Power.	Reduced by Test.	Loss per Cent.	Maximum Power.	Reduced by Test.	Loss per Cent.
Hard Bars · ·	H	38·20	33·15	—	31·50	28·15	13·5	27·13	23·54	13·8	14·15	1·5	92·5	10·0	9·8	8·5
Tempered throughout.	T	31·50	22·30	33·3	27·10	18·0	36·6	22·0	9·43	57·7	17·5	6·16	64·1	10·48	2·3	81·1
Tempered at the ends.	E	24·30	10·22	59·9	19·50 / 27·15	6·32 / 16·52	68·1 / 41·1	18·13	3·32	82·2	14·40	3·6	79·5	11·5	1·3	90·8
Soft Bars · · ·	S	18·50	3·37	81·5	16·32	0·45	95·6	15·8	-1·35	110·4	11·30	-3·22	129·1	9·13	-2·50	130·0

Here we find, as pointed out when speaking above of the principle of tenaciousness, the striking correspondency of the tenaciousness with the hardness, extending, with but one exception, throughout the whole of the series' of bars. Without correcting the results for the differences of weight in the bars designed to be of the same dimensions and mass—which, as to the effect of the differences on the property under consideration, were comparatively inconsiderable—we adduce the set of bars No. III., being of the average weight of 720 grains, as a striking example of the correspondency of the ratio of tenaciousness with that of the hardness as described by the manufacturer. And these four bars, taken in the order of their hardness, commencing with the hardest, afford, we perceive, this series under the head "loss per cent.," as the measure of the tenaciousness, $i.\,e.$ 13·8, 57·7, 82·2, 110·4.

So conclusive, I conceive, are the results thus elicited, as to afford some encouragement to expect, that, by this process, not merely the order, as to hardness, of a series of similar bars or plates of steel, may be determined; but, possibly, an absolute scale and standard of temper (as illustrated more particularly further on) may be eventually obtained.

But the foregoing investigations, with the

quadruple series of bars of different thicknesses, afford, in the observable augmentation of power, or capacity for the magnetic condition, with each modicum of increase in the degree of hardness, another mode of testing the temper of bars, if beyond a given minimum of thickness.

Suppose the bars to be tried, for instance, to be of the same length, and similar in the quality of steel, as those employed in the present researches; and let the weight of any particular bar, to be tested for temper, be 1400 grains, and its extreme power, when magnetized, as expressed by its action on the compass at two lengths' distance, a deviation of 27° 34′, the tangent of which is 522. Now this tangent being set off as an ordinate, from the weight 1400 grains, in the line of abscisses in the diagram, Pl. ii., will extend a small distance above the height of the curve T, belonging to bars tempered at a blue throughout—indicating that the bar is a very little harder than that temper.

A similar approximation of the degree of hardness, in steel of a like quality, will also, by means of the diagram, be afforded, for bars of other lengths. Suppose, for example, a bar of 7·5 inches long, weighing 5000 grains—the action of which at the distance of 15 inches, or two lengths', from the compass, produces a deviation, say, of 28° 22′, of which the tangent is 540.

As similar bars, or bars of proportional dimensions, are, to each other, as the cubes of their
lengths, respectively—the weight of this bar,
compared with its equivalent of six inches long,
will be as $7\cdot5^3$: $6\cdot0^3$: : 5000 : 2560 grains.

Hence the place of this bar, on the line of
abscisses, is at 2560 grains, from whence the
tangent of its power being projected as an
ordinate (see diagram, Plate II.) will be found
to reach more than half way betwixt the second
and third curves, to the point s'; shewing that
the degree of hardness is less than that of spring
temper throughout, but greater than that of
bars hardened only at the ends.

Now, if the ordinate due to the 7·5 inch bar
were prolonged to h, as far as the upper curve
(that indicative of the power of perfectly hard
bars), then the relation of the limited portion of
the ordinate of this particular bar extending
above the lower curve, to the complete portion
of the extended ordinate comprised betwixt the
upper and lower curves,—the curves belonging
to steel in extreme conditions as to hardness—
would afford a kind of measure of the degree of
hardness of the particular bar.   For, if we call
the distance betwixt the extreme curves in the
diagram, 100°,—that is, the temper of the soft
bar $S = 0$, or zero, and that of the perfectly
hard bar $H = 100°$,—then the temper of any

intermediate bars, of the like mass, might be specified, in the proportions of their ordinates rising above the lowest curve, S, respectively, to the total interval of 100°. Hence, in the case of the particular bar under consideration, its place in the ordinate would be at s'—which, according to simple proportion, reckoning the whole interval betwixt s and $h = 100°$, would be about 50° of such extent,—and the measure, proximately, of its hardness, therefore, would be 50°.

It is not presumed, however, that the measure of hardness, thus indicated, would be in the true proportion of its number of degrees—for the relations of the several curves vary in different extents of the abscisses,* whilst it has not

* Though the simple relations of the curves of temper vary so much in the beginning of the series, yet if we combine the two qualities, as expressed by the tangents of the power of maximum capacity and that of tenaciousness (or power remaining after the application of the test-bar), the proportions, then, of the relative powers of different degrees of hardness, make rather a striking advance toward similarity. Thus, for instance, comparing the two series of bars, No. II. and V., as exhibited in the re-modelled table given in this chapter,—the former weighing, on an average, 145½ grains, and the latter 165 grains, — the several sums of their individual tangents of maximum powers and reduced powers, are found to be as follow:—

Sum of Tangents of Powers.		Reduced to the same Proportions	
Bars, No. V.	Bars, No. II.	Bars, No. V.	Bars, No. II.
337	1158	100	100
227	838	67	72
214	647	64	56
113	310	33	27

been proved that the power of sustaining the magnetic condition proceeds regularly in the ratio of the hardness.

The relative positions of the extreme curves, moreover, are only to be considered as proximately correct, for bars hardened, and softened, in the ordinary way; for the untempered bars, by very slow cooling, or other peculiar process, might have been made to yield a curve very far below that marked S, in the diagram.

Yet, within certain limits, as to the mass of the magnets, we might thus obtain a kind of *sclerotesmetrical* scale available for many important objects, not merely in the construction of magnetical instruments, but extensively useful in certain manufactures of steel. For, for the testing and comparing of large numbers of needles or plates designed for compasses, constructed out of steel of similar quality; or for proof and selection in instruments of the same form and description—the powers of the steel made use of might be easily obtained and projected in a diagram, so as, within the small comparative differences of weight in articles of any particular size, to give a really satisfactory scale of temper. In such case it might be advisable to correct the curves of temper for the equation of the tangents, in order to obtain a more equable scale—the proper correction, of course, being also applied from a table of equa-

tions to the power of each bar, plate, or instrument, as given proximately by the tangent of
its deviation.

The foregoing observations on the testing of
temper, apply, mainly, as is indicated in the
title of this chapter, to instruments, etc. constructed of the *same kind* of steel. It is not
improbable, however, but both test and scale
might prove to be available for those different
kinds of steel, as to denomination, which are
regularly manufactured out of the same description of iron; though for steel which has been
made of iron of other qualities, the scale would,
no doubt, require, as the investigations of the
next chapter will shew, to be modified.

It would have been desirable, had I had time
and opportunity for carrying on these investigations, to have ascertained, by a course of
experiments corresponding with those of which
the results are exhibited in the diagram, the
relative powers of bars in similar series' constructed of steel of the different denominations
and qualities usually known in commerce. But
my professional duties prevent me undertaking
an inquiry so extensive. Such investigation I
must therefore leave for any one who may have
interest enough in this department of magnetical
science, and who may be able to command the
leisure necessary, for pursuing the subject.

CHAPTER VII.

THE DETERMINATION OF THE *QUALITY* OF BARS, PLATES, OR INSTRUMENTS CONSTRUCTED OF STEEL.

———

THE principles of inductive science lead directly to the conclusion, that the sustaining power of the magnetic condition in a mass of steel, should afford the means of determining, as to the degree of carbonization at least, the *quality* of such steel. For as simple uncarbonized iron, notwithstanding its high capacity for magnetism, possesses extremely little power of retention, and as thoroughly carbonized iron, with no higher, but rather a lower, capacity, has a very great capability for permanent magnetism—the inference becomes obvious, and the conclusion, apparently, inevitable, that a *partial degree* of carbonization must be attended with a limited measure of sustaining energy : and if so, then, a philosophical process for determining, in some degree, the extent of carbonization in different qualities of steel, is at once suggested, — viz., by ascertaining, as we find can be done so easily, either their relative tenaciousness, or their capacity for permanent magnetism, when in a certain known condition as to temper. For either of these processes are available, except in the case of thin plates, for this object; because the extent of carbonization, and the degree of hardness, alike tend to give increase of power for sustaining the violence of the magnetic condition. Hence in bars of the same degree of hardness— such as thoroughly soft, or as hard as possible, being conditions in which the greatest similarity may be obtained—a most observable agreement is, by actual experiment, found to subsist betwixt their capacities for magnetic energy, and their power of retention, whilst a similar correspondency seems to exist betwixt these two properties and the degree of carbonization.

This subject, however, may be the better illustrated—both as to the general availableness of the method suggested, and the extent of reliance which may be placed on the results—by applying the principle, practically, to different specific conditions of the steel to be tested.

I.—*For the determination of the quality of the unmanufactured, or raw material.*

The results obtained in investigating the magnetical properties of steel of different deno-

minations in commerce — as to capacity and
retentiveness—will at once serve to verify the
principle, and to explain the process of its application for the purpose specified.

For this investigation, I procured, in the first
instance, duplicate sets of bars of the form and
dimensions of those before employed, and denominated No. I. : viz., bars of a rectangular form,
6 inches long, and 0·5 inch square—of the several
denominations of " best cast steel," " best shear
steel," and " common bar steel," unrefined.
These were ordered to be made as hard as possible; but, from the action of the glazing wheel,
probably, in polishing them, their hardness was
found, on examination by the foregoing test, to
be reduced very nearly to that of a spring temper. The results afforded by them were not
altogether satisfactory; for although a most
marked distinction, in magnetical properties,
was found betwixt the two best denominations
and that of " common bar steel,"—yet betwixt
the " cast steel " and the " shear steel," the
differences were so inconsiderable as to suggest
a doubt of their being characteristic specimens
of these best qualities.

Another set of bars of a like mass and form,
nearly, was, therefore, procured, of qualities
which, in commerce, are known to be more
diverse : viz., " spring-steel," (employed for the

springs of carriages) two qualities of " shear steel," " blister " or " unrefined steel," and ordinary " cast steel."

These bars proved to be of a length varying from 5·95 to 6·07 inches, and from 0·52 to 0·55 inch square. Their weights, with the marks by which they are hereafter discriminated, were as follow:

Description of the steel.	Discriminating mark	Weight in grains.
Common Spring.	*S P.*	3380
Single Shear.	*S S.*	3280
Double Shear.	*D S.*	3390
Blister.	*B.*	3360
Common Cast.	*C C.*	3550

All these were first examined in the state, as nearly as possible, of the raw material—except the subjecting of them to a like red heat and letting them cool slowly on the hearth of the forge—that the degree of temper, or softness, might be uniform. The power of each bar, after being magnetized to the extent of its capacity in this soft state, was tried at two lengths' distance from the compass; and again, after the whole had been in succession applied to the action of the test-bar—when the results, in both ways, were found to exhibit a considerable correspondence with the order of their qualities, as regarded in commerce.

The same bars were then hardened at a bright
red heat, by being plunged, when in that condition, into a pail of cold water, holding a
quantity of salt in solution. Their powers, both
as to capacity and retentiveness, were again
examined on the trial-board, as before. The
hardness, however, not having been found to be
such as was desired, they were again subjected
to the process at a *white heat*, and in that state
— which was found to be quite satisfactory—
their magnetical powers were also determined.

The results of the first and the last experiments — in which the equality of temper, and
general approximation to the extremes of hardness and softness, were the most satisfactory—
are contained in the following table

Discriminating mark.	Highest Power.	Reduced by Test Bar H.*	Tangents of Maxima and Reduced Powers.					
			Maxima.	Reduced.	Difference.	Loss per Cent.	Sum.	Ratio.
I.	II.	III.	IV.	V.	VI.	VII.	VIII.	IX.
			I.—Bars made quite soft.					
SP.	7·25	−0·10	127	− 3	130	102	124	30
SS.	11·54	+0·45	211	+ 13	198	94	224	55
DS.	15·7	2·22	270	41	229	85	311	76
B.	15·43	3·0	281	52	229	82	333	82
CC.	16·52	5·57	303	104	199	66	407	100
			II.—Hardened at a White Heat in Salt Water.					
SP.	26·10	16·3	491	288	203	41	779	47
SS.	36·15	31·15	733	607	126	17	1340	82
DS.	37·15	31·55	760	623	137	18	1383	85
B.	36·52	32·30	750	637	113	15	1387	85
CC.	41.10	37·20	874	763	111	13	1637	100

* The deviating power of the Test Bar at the commencement of the
experiments, was 34° 20′; at the conclusion it was found to be 32° 0′.

In these results, a considerable relation is observable betwixt the magnetical properties of the several bars and the respective qualities of their denomination of steel (any exceptions being owing, not improbably, to the want of conformity, in the steel employed, to its proper characteristic); and what is still more striking, the similarity of the results indicating the relative qualities (however differing in the proportions of such qualities), as alike yielded by the magnetic capacities of the bars when soft and when hard (col. iv.), as well as by their tenaciousness for the magnetic condition, under the action of the test-bar, in each of these states, as to temper, as shewn by the loss per cent., given in col. vii. Results, also, similarly accordant, are obtained by combining the power as to capacity (col. iv.) with that of reduced powers, by the test-bar (col. v.), which is here done (somewhat empirically indeed), col. viii. by simply taking the sum of the tangents of the maxima and reduced powers.

This experiment, however, is not by any means to be considered as conclusive, or such as to afford decided results, as to the extent of applicability of the magnetical test, for the determination of the various differences in the quality of steel. For I am not at all assured that the several kinds of steel employed in this experiment were characteristic specimens ; neither have I any knowledge of the quality of the iron out of which the steel was "converted."

Hence, while nothing more than broad and general inferences can be expected from such an experiment; yet some such general conclusions, I conceive, may be obtained. For so far it is satisfactorily shewn, that certain differences in the quality of steel, affect, in a most striking manner, the magnetical properties of that substance ; and that these differences in quality and magnetical properties, have such a relation, that the magnetical properties *may* be available, not merely for ascertaining the degrees of carbonization—as they manifestly are,—but, not improbably, for the determination of the essential quality of the steel, with relation to the quality of iron out of which it may have been manufactured. And, perhaps, it is not unreasonable to expect, that, were all the varieties of magnetic capacity in each denomination of steel, and in each quality, as respects the kind of iron out of which the steel is "converted," experimentally ascertained—that a strictly scientific process of testing, founded on these principles, might be devised; a process which might, possibly, exhibit results, if not as exact, at least as conclusive, in certain most important relations of value in the metal—as are obtained by the beautiful process of

assaying. And if so, even in a moderate degree
of perfection only, it needeth not to say how
important a contribution it might prove toward
this great desideratum in the arts.

Subsequent experiments, which to a considerable extent I have made, have served still
further to support the general views above
stated; but they have not been pursued sufficiently far for the determination of the precise
effects of the various differences in the methods
of manufacture of steel, or modes of tempering
it, by which its magnetic properties may be
modified.

Preparation, indeed, for the particular investigation of the effects of some of these differences
in the magnetic qualities of steel, had been made.
For this, Messrs. Sanderson and Co. of Sheffield,
and Messrs. Stubs and Co. of Warrington,
kindly gave me their assistance, by supplying
me with specimens of steel of their respective
manufactures, comprising a considerable variety
of kinds, converted out of the best qualities of
foreign iron, and with some differences in the
degree of carbonization. Further experiments,
however, were found to be necessary, and with
a still larger variety of specimens, before the
particular points contemplated in the investigation could be determined. But, as to the general

relation betwixt the qualities of the steel and their respective magnetic capabilities, the expected results were most satisfactorily realized. Certain peculiarities, arising from different *modes* of hardening, and others belonging to some of the characteristic differences in the process of manufacture in the steel (as indicated by their denominations in commerce), will come under consideration in another part of this volume, where the various alterations in principle or construction, suggested by these investigations for improving the quality of magnetical apparatus, are described.

II.—*Applicability of the principle, now proposed for testing the quality of steel, to instruments, and other manufactured articles.*

This principle of testing the *quality* of steel, as to its more characteristic differences, is even now (to a certain extent) applicable to manufactured articles of particular kinds. For if the manufactured articles be of a regular form, with parallel sides—such as the plates for surgeons' saws, as well as saws for some other particular purposes; as also the bars or plates designed for sea compasses, dipping needles, or other important magnetical instruments—there would be no great difficulty in determining, by the trial of their magnetical properties, the relative goodness, to a decidedly important measure of minuteness, of the steel of which the several articles might be composed.

Of the practicability of accomplishing this, with relation to the degree of carbonization, we have decided proof, I conceive, in the Navy compass needles, whose different magnetical properties are described in Chap. IV. For in the case of needle No. vii., we find its maximum power to be so greatly below that due to a corresponding mass of good cast steel, that its quality is manifestly very inferior. Assuming its temper, therefore, to be the very lowest possible, or perfectly soft, we should infer, by observing its place in the diagram, Plate ii., [NC∙VII] that it had not much more than a third of its proper dose of carbon—being little superior in quality, indeed, to common iron !

In another instance, a curious and interesting verification of the applicability of the magnetical test, for the determination of the quality of steel, unexpectedly occurred, whilst a subject of altogether a different character was under investigation. For the comparing of the magnetical powers of solid bars of a given mass, with those of an equal mass of hardened steel plates of corresponding form and length, I had ordered, of a clever manufacturer of machinery at Keighley, the bars of steel requisite for the experiment.

These consisted of bars of different thicknesses, but otherwise alike,— the order being, "that they should be made out of the same steel, and tempered exactly in the same way." On magnetizing the bars, and testing their powers, I found, to my great disappointment, that the quality of the thick bars was altogether different from, and inferior to, that of the thinner.

The bars, in the first instance, having been tempered at the ends only, whilst left soft in the middle, as I had ordered,—I returned them to be prepared for another trial, and for comparison with their former state and with each other, by being made quite hard throughout. This was done, and the difference of *quality* in their magnetic state was found similar to what it was before — the thicker bars being very inferior.

On this verification of my first impression, I wrote to the manufacturer, remonstrating with him for the deviation from my instructions, intimating, that he must have made the bars out of *different qualities* of steel, and informing him of the proof I had obtained. His reply proved alike satisfactory and interesting. He stated that my orders had been strictly followed; that all the magnets had been made out of the same bar of *best blister steel*, which was of *the proper size for the thicker bars;* "but," he added, "having had to form the thinner magnets by drawing out the large massive bar, the great deal of hammering to which they had been subjected must have improved the quality of the metal!"

And this was no doubt the fact; — a fact beautifully verifying the soundness of the principle of the *testing*, herein suggested, for the determination of the quality of the steel; supporting also the supposition of an essential relation existing betwixt the quality of steel and that of the iron out of which it is converted; as well as approving the practice by which the manufacturer seeks to improve the quality of cast steel by the process of, what is termed by the workmen, *shearing*.*

The improvement in the quality of the steel,. in the bars above described, was so considerable, that those of a quarter of an inch in thickness only, were found to have very nearly as high a degree of magnetic energy as those of twice the thickness.

The magnetical test is found to be further applicable, on the same principle, to the impor-

---

* The reason why *shear steel* should not be preferable for magnets, in certain cases, or in general practice, to cast steel, when the tendency of the process is to improve the quality, admits of a simple explanation, as hereafter illustrated.

tant practical object of determining any other changes, which may occur in the quality of steel, whilst under the hand of the artisan or manufacturer—such as, for instance, the deterioration which sometimes takes place, by the steel being overheated, either in the process of welding or hardening.

By means of this test, I found, that the quality of the metal, in several sets of highly tempered thin steel plates, which I ordered for the construction of compound compass needles, had been greatly deteriorated by mismanagement in hardening,—assuming, as I was assured by the manufacturer, that they had been made of a specified quality of steel. A measure of decarbonization appeared, in these cases, to have taken place,—the quality proving so inferior as to render the plates entirely useless for the object designed.

Some measure of actual deterioration in quality, by repeated heating and hardening, was shewn in the case of a set of steel bars expressly appropriated for the trial. The repetitions of the process were both numerous and severe. The resulting condition of the bars, as tested by the method here suggested, though not so inferior to what it was at the commencement as had been expected, was yet, I think, sufficiently decisive of a measure of deterioration having taken place.

Treating here chiefly of the principles on which the quality, and degree of hardness, of steel are, as it is I think clear, in a certain measure determinable, we defer the details of the practical processes to their proper place in a subsequent part of this volume Our present object has been just to illustrate the mode in which, it is conceived, the desirable object contemplated may be accomplished.

CHAPTER VIII.

INFERENCES, FROM THESE INVESTIGATIONS, CONCERNING THE
PHENOMENA OF MAGNETISM IN STEEL, AND ON THE
APPLICABILITY OF THE RESULTS TO THE IMPROVEMENT
OF MAGNETICAL INSTRUMENTS.

THESE investigations, to the extent developed in
the preceding chapters, give us a glimpse of the
rationale of the magnetic condition in a mass
of steel. At the same time they fully support
those views by which I was led to adopt the
principle, in the first instance, of a *thorough*
tempering of the plates or bars designed for
compasses or magnets, as essential for any real
improvement in the instruments at present in
use amongst us; and, ultimately, of making the
steel *quite hard throughout*, in order not only to
the attainment of the greatest possible permanency of the magnetic power, but also of still
higher qualities, as to the energy of the magnetic condition, in such instruments.

The neutral state of the magnetisms in ferruginous bodies is evidently the natural condition,
and the condition of polarity one of violence.
Hence, in iron, which has very little sustaining

power, the magnetic condition, however highly developed, is verÿ evanescent; and in steel, unless hardened, the power is not permanent. But in hardened steel, the degree of tenaciousness, or permanency, increases according to the degree of hardness. This permanency might be supposed to be maintained by hardening in either of two ways—by the hardness of the ends *suspending*, as it were, the magnetic tension (the theory which seems to have formed the basis of the system on which artificial magnets are usually constructed); or by the hardness throughout the mass *supporting*, as by a rigid pile, the accumulating force of the *magnetic elements*, from particle to particle, to the extremities.

That this latter was the fact, I had long ago ground for inferring, from having tried to construct a large bar magnet (three feet long, and about three inches broad) out of a flat bar of iron, steeled and tempered only at the ends. But it was found to receive an exceedingly weak power. In truth, there was no supporting rigidity in the iron, and, therefore, no means of maintaining the violence of the magnetic pile—or what may be called, *battery* of magnetical elements—so that the moment the developing power was removed, the magnetisms, for the most part, returned to their neutral condition.

The facts developed in this division of my magnetical investigations, are instructively accordant with this view. For we have seen, in the results given in Chapter V., that, however comparatively energetic thin or small magnets may be if hardened only very partially—massive magnets, so tempered, soon approach a maximum power, beyond which, whatever augmentation may be made to the mass, little or no addition is yielded to the power of the magnet; obviously because the state of the steel has not hardness and rigidness to sustain the force always tending towards the neutral condition. Hence, we perceive how, by increasing the degree of hardness, we obtain strength or rigidity for sustaining a greater degree of violence, and so have the means of employing larger masses with advantage, by which more powerful magnets may be obtained. And, by analogy, we should infer, that alloys of steel, in which an increase of hardness is attainable, may, not improbably, afford in a still higher degree the peculiar quality of fixedness, so as to yield artificial magnets of the very highest powers.

These views, concerning the phenomena of the magnetic condition, as they are exhibited in ferruginous substances, being verified, not only by all the investigations hitherto given, but by a large variety of others yet to be described, may fairly be assumed, I trust, to be, in principle at least, sound in science. Nor are the *principles* on which it is proposed to test the *quality*, as well as the *temper*, of steel, incapable of being maintained by similar sanctions. For the supposition of a physical relation existing betwixt those qualities of steel which constitute its characteristic excellence, and the magnetic capacity of this substance, is by no means an arbitrary assumption. Good reason, indeed, can easily be given, why, on scientific principles, such a relation should exist; and on these grounds:—

If the matter of iron have, in a supereminent degree, the highest inherent capacity for magnetism among all known substances—then it should be expected, that the specimens of this metal, *most perfectly iron* (or the most free from admixture with other substances) should have the highest magnetical properties among the manifold varieties of this metal. And this— as in a paper in the "New Edinburgh Philosophical Journal," October 1822, p. 272, I have already intimated and partially illustrated—I doubt not to be the fact. For in the investigations there referred to, I found a constant relation betwixt the *ductility* of iron (a property, in metals in general, belonging, in the highest

degree, to the purest qualities), and its magnetic
capacity. Hence, from analogy, it was equally
to be expected, that the steel of any particular
kind, as to process of manufacture, which had
been formed out of the purest iron, should possess the highest magnetic qualities of its particular denomination. That is, if the *best iron*
possesses, because of its purity, the highest magnetical quality belonging to iron; the *best steel*—
which, in any particular denomination of manufacture, is well known to be yielded by the best
iron—should be expected to have the highest
magnetical quality belonging to steel.

This view of the kindred properties of iron
and steel, in their capabilities for the magnetic
condition, does not profess to apply to steel
which has been manufactured either out of the
same, or diverse qualities of iron, by *different*
processes; but only to steel which has been
"converted" out of different qualities of iron by
the *same* process. As to the magnetical relations with one another of the different denominations of steel, generally—such as those of
"cast steel," "shear steel," and "blister steel"
—there is an apparent diversity, besides what
belongs to differences in the quality of their
respective metal, as we shall hereafter have
opportunity of shewing; but as to the magnetical properties of any particular denomination of

steel of different qualities, both the principles
above submitted, and the experiments described
in the foregoing chapters, agree in justifying
the inference, that such different qualities of
steel should exhibit magnetical properties with
a special relation to the qualities of their original
iron respectively. The *cast steel*, for example,
converted out of the best Swedish iron (such as
that known technically, with reference to its
mark, as *hoop L*), would be expected to possess
higher magnetical properties than the cast steel
made out of foreign iron of acknowledged inferiority of quality; whilst the varieties of cast
steel from these inferior qualities, would, if our
principle be correct, have magnetic qualities
diminishing in their powers according to the
order of deterioration in the several qualities of
their original iron.

So far, then, as these relations go—with reference to different steels of *the same denomination*,
as also, with reference to steels of different degrees of carbonization — the principle which has
been suggested for testing *the quality*, may, I
presume, be held to be sound and philosophical.
But as the multifarious relations betwixt the
magnetical properties of steel of *different denominations* and their varying qualities, are not
known—I do not venture to determine whether
the embarrassment necessarily produced by

varieties in quality and varieties in process of manufacture, may, or may not, be decidedly overcome. Yet, I cannot but suppose it quite possible, that these relations, if thoroughly investigated — as to capacity, or retentiveness, in different conditions as to hardness, or as to capacity for induced magnetism (the test employed for ascertaining the quality of iron), or as to the different relations of these several properties in different qualities and denominations of steel — might afford such discriminating peculiarities or differences, as to be rendered practically available for the general purpose of testing.

The views suggested at the beginning of this chapter, concerning the phenomena of magnetism in steel, apply directly to the subject of practical magnetics, and yield important guidance for the improvement of magnetical instruments. They shew us the fallacy of certain principles, heretofore accustomed to be acted upon, in the construction of such instruments— such as the supposed inferiority, in magnetic capacity, of very hard steel — the imagined benefit of reducing, by tempering, the preparatory hardening — the asserted inferiority of cast steel, and the supposed indifference as to the quality of the steel employed for magnets,— whilst the results of the foregoing investigations determine, and that most satisfactorily, certain definite and intelligible principles on which an improved quality of apparatus may be obtained.

For the results now derived from the preceding investigations have suggested the construction of *compass needles* and *variation needles* (referred to in Chapter I.) on the *compound principle*, instead of that of single bars—as in that particular construction, besides other advantages, there is a very beneficial attenuation of the degree of magnetic concentration belonging to a similar mass in a single bar; such an attenuation, that the violence on the sustaining power of the plates, tending to deteriorate the ultimate effect of powerful combinations, is greatly diminished.

The application of the results, moreover, derived from the foregoing investigations, enable us to perceive in what respects all kinds of magnetical instruments may be improved — suggesting, as principles for obtaining such improvement, the selection, by the method of testing, of the most powerful and tenacious plates or bars; the obtaining of stronger, and almost unchangeable, instruments, by using the hardest, instead of soft or tempered steel; the further improvement of these by employing steel of the finest and most valuable qualities, such as those converted out of the best qualities of iron; and the gaining of still higher energy, and increased permanency, in powerful instruments, by the combination of numerous thin plates, in place of the usual massive bars, and, where applicable, of *separated* thin plates, for the greater diffusion of energy, and diminution of the tension, of the magnetic force in the instruments.

These principles, derived by induction from an extensive series of original investigations, I applied to practical purposes some years ago, and, it is believed, with the most satisfactory results. As to a general summary of them, the results were given in the first edition of this part of the present work, published in the month of March, 1839; but as the practical arrangements, which experience has suggested, for the construction of these improved instruments and apparatus, as well as the details of researches for determining *the laws of combination* in magnetised steel plates, etc., have their place in the subsequent divisions of these Magnetical Investigations, — this notice of the general advantage obtained by the principles of construction developed and illustrated, in the foregoing pages, may here suffice.

MAGNETICAL

INVESTIGATIONS.

BY THE

REV. WILLIAM SCORESBY, D.D.

FELLOW OF THE ROYAL SOCIETIES OF LONDON AND EDINBURGH;
CORRESPONDING MEMBER OF THE INSTITUTE OF FRANCE,
ETC. ETC.

PART II.

COMPRISING INVESTIGATIONS CONCERNING THE LAWS OR
PRINCIPLES AFFECTING THE POWER OF MAGNETIC STEEL PLATES
OR BARS IN COMBINATION, AS WELL AS SINGLY,
UNDER VARIOUS CONDITIONS AS TO MASS, HARDNESS, QUALITY,
FORM, ETC., AS ALSO CONCERNING THE COMPARATIVE
POWERS OF CAST IRON.

LONDON:

LONGMAN, BROWN, GREEN, AND LONGMANS,

PATERNOSTER-ROW.

M.DCCC.XLIII.

LONDON:

PRINTED BY MANNING AND MASON, IVY LANE,
PATERNOSTER ROW.

ADVERTISEMENT

TO THE

SECOND PART

OF THE

MAGNETICAL INVESTIGATIONS.

———

THE experiments described in this second Part of "Magnetical Investigations," were, to a very considerable extent, made some years ago. They have been carried forward, however, at intervals, as the Author's arduous professional duties would admit, to the present time; and the publication of the results obtained has been deferred until all the objects of inquiry, belonging to the department of experimental research, which were originally contemplated, or have been suggested in the progress of the Investigations, have been completed.

It is hardly needful to say that the researches have been of a very elaborate nature and extent. The Tables given in this part of the work alone are the result of above four thousand observations

on the deviations produced by the numerous
magnetical plates and bars subjected to experiment,—each observation requiring the needle
of a five-inch compass, after being disturbed by
the influence of the magnet to be tested, to attain
a stationary position, and the angle of deviation
from the magnetical meridian, to be read off to
within a minute or two of a degree. And besides
this labour by the method of deviations, a large
number of magnetical bars of the horse-shoe form,
etc., had to be otherwise tested ; and the results
obtained by the different modes of experiment
required to be tabulated, in many cases at a
considerable addition of trouble, for reducing the
observations, and for obtaining the exact measures
and weights of the bars or plates employed.

Similar investigations, as some of these here
described, have, as is well known, been before
made, and analogous results, in certain cases,
obtained. The very extensive range of inquiry
here pursued, however, with a constant adher-
ence to the same modes of experimenting and
testing, will, the Author hopes, not only excuse
his having gone over some ground previously
examined ; but will yield a measure of newness
—by the unity of method and ampleness of investigation in the vast varieties of mass, form,
quality, temper, and denomination of the magnets made use of—to the researches themselves.

Besides the experimental inquiries of the first
and second Parts of these Magnetical Investigations, the Author has always contemplated, if
leisure and health should, in Providence, be
yielded him, to extend the work to two or
three additional parts, embracing,—Practical
Magnetics, with the application of these Investigations to the improvement of Magnetical
Instruments and Apparatus; various original
illustrations of, and experiments on, Magnetical
Principles, and Phenomena, etc., etc.

It may be proper to state, however, that the
two Parts of the work now before the public
have been so adjusted as to be complete in
themselves,—the last sheet of Part I., having
been cancelled, and the substituted sheet G
placed at the conclusion of this Part, in order
that the connection betwixt the two portions
might be more perfect.

The Author has only, in conclusion, to add,
that whatever benefit to science his labours
might be calculated to yield, or to his country,—
for aiding the improvement of an instrument
heretofore so exceedingly defective and uncertain as the sea-compass,—such benefit he is most
desirous should be fully realized, and he has
taken some pains to render that desire effective.
In order to this, the Author has communicated,
without reserve, to Messrs. Stubs (of the firm

" Peter Stubs "), manufacturers in steel, of
Warrington,—the principal practical modes of
constructing, magnetizing and testing the various
kinds of plates and bars described in this work,
or designed for being employed in magnetical
apparatus.   The employment by Messrs. Stubs
of first-rate workmen in all the departments requisite for the construction of magnetical instruments, and especially for hardening or tempering;
their extensive engagement as manufacturers of
steel; with the deserved celebrity which their
files and tools have gained for their firm, not
only in England, but in many parts of the world
abroad,—will at once enable them to undertake
the kind of work referred to with peculiar
advantages, and be a guarantee to the Author
and to the Public, of its being faithfully and
efficiently done.

BRADFORD, YORKSHIRE,
November 7th, 1843.

CONTENTS.

PART II.

CHAPTER IV.

ON THE RELATIVE POWERS, IN COMBINATION AND SEPA-
RATELY, OF HARD PLATES OR BARS OF STEEL OF
DIFFERENT DENOMINATIONS AND QUALITIES - - 185

CHAPTER V.

ON THE MAGNETICAL POWERS OF STEEL PLATES OF
DIFFERENT MASSES, DENOMINATIONS, AND QUALITIES,
AND MEASURES OF COMBINATION, PREPARED FOR
THE NEEDLES OF SEA-COMPASSES, WITH THE EFFECT
OF SPACING THE PLATES. - - - - - 202

CHAPTER VI.

CHAPTER VII.

CHAPTER VIII.

OF THE MAGNETICAL POWERS, RECEPTIVE AND PERMA-
NENT, OF *CAST IRON*, BOTH IN SEPARATE BARS
OR PLATES, AND IN VARIOUS COMBINATIONS OF

CHAPTER IX.

ON THE MEASURE OF PERMANENCY OF THE ENERGY
IN STRAIGHT-BAR MAGNETS, BOTH SINGLE AND
COMPOUND, AND OF DIFFERENT DEGREES OF HARD-
NESS, AS SELF-SUSTAINED, OR AS INFLUENCED ONLY

MAGNETICAL INVESTIGATIONS.

PART II.

CHAPTER I.

AS TO THE POWERS OF COMBINATIONS OF MAGNETISED PLATES OF TEMPERED STEEL, IN CONTACT.

THIS important object of inquiry—as affecting both the directive power of compass needles and the energy of compound magnets--has been pursued to a much greater extent, and with a much larger variety and assortment of plates and bars, than it may here be necessary, or indeed useful, to describe.

In the whole, the powers in combination of about forty sets of plates or bars, have been carefully tried, in their respective series of from 2 to 192 together,—the total amount of magnetised pieces of steel, subjected to experiment, being scarcely less than from seven to eight hundred. And in several of the sets of bars or plates, the powers in combination were likewise

determined for every amount of numbers, by
successive additions of bar by bar, or plate by
plate, up to the total quantity in the particular
series.

The numerous series employed were of various
dimensions, from plates or bars of six inches, to
two or even three feet, in length. They comprised steel of different qualities, and of the
several conditions of hardness, from that which
was perfectly soft to that which was "as hard"
---to use the technical language of the smith or
cutler—"as fire and water could make it."

The investigations concerning the relative
powers in *combinations* of bars and plates of different weights, served to confirm the principles
deduced from the experiments made with *single
bars* in Part I. By applying the powers, indicated by the tangents of the deviations, to the
diagram, Plate II., and correcting them by the
scale of equation of deviations, the combinations
of magnetic plates, as compared with the powers
of single bars of the *same length*, were satisfactorily determined. The comparison of bars of
different lengths, however, by this process, is,
unless proximately, a matter of some difficulty.

A selection of the most characteristic and
conclusive of these investigations will now be
given. But, in this chapter, the selection will be
only of a particular class of experiments, made

with steel tempered in the same way, and of
a similar degree of hardness. The resulting
temper of these bars and plates was that of a
spring temper *throughout* the mass,—each bar
or plate being first made quite hard, from end
to end, and then reduced to a blue colour, by
a second heating, and there fixed.

As the results with steel plates of different
lengths were found, in all essential points,
correspondent—one example may be sufficient
to be described in detail. And for this it will
be convenient, for the purpose of comparison,
to adopt the experiments made with the two-
feet plates of tempered cast steel described in
page 38. These plates, it may be remembered,
constituted an uniform series of thirty-two in
number—being 1·5 inches broad, and ·042
inch thick, and of the average weight of 2869
grains. The quality and temper of these plates
were in general very equable; the deviating
power on the compass, at one length distance,
being, in the weakest, about 15°, and, in the
strongest, 18° 30′, and the mean power of the
whole, as tried separately, being 16° 10′.

These plates were all pierced with four corresponding holes, two about an inch from the
centre, and two about an inch and a half from
the ends, as represented in the annexed figure:

By these holes, any number of the plates could be readily combined together, and firmly screwed by means of brass pins, with screw nuts, put through them. The particular size of the plates was adopted for the purpose of a more satisfactory comparison with a pair of bars by M. Gauss, in the Royal Observatory at Greenwich, which appear to be of admirable quality and of very superior power. And this degree and mode of tempering was fixed on, at the early stage of my researches, as the first advance towards an improvement in the ordinary method of constructing compass needles and magnets, in this country,— the prevalent method consisting of a mere tempering of the ends, whilst the intermediate portions are either left in the original state of the steel, or, according to Capt. Kater's recommendation, *reduced* to the soft state. For even this alteration, in the usual method of constructing magnets, it was inferred, as the results already described abundantly indicate, would yield additional retentiveness ; and, by virtue of that quality, would admit of a larger number of plates or bars being combined with advantage, than when tempered, after the ordinary method, by a moderate hardening only at the ends.

Corresponding with the steel plates now described—in length, breadth, and quality of steel —I procured from the same manufacturer four bars of about ·62 inch in thickness,—the two best of these being of a similar degree of hardness with that of the two-feet plates, and weighing 6·281 and 6·236 pounds of 7000 grains, or 6·258 pounds, the mean weight.

The modes of magnetising the plates and bars, and of determining their respective powers having been described in Part I. Chapters II. and III,—we are prepared for the consideration of the investigations, with this apparatus, concerning *the law of combination of magnetised plates of tempered steel, in immediate contact.*

The whole series of two-feet plates was, in the first instance, carefully magnetised to their utmost capabilities, by means of the two-feet bars above described, and by the process K.S. (p. 13), and their power of deviation determined at the distance of two feet, or one length, from the centre of the compass, on the " trial-board " (p. 31). One fasciculus was then formed of the whole of the plates, and the position of the plates changed by separating the fasciculus about the middle, and replacing the two portions in a different manner ; so that those at first in the middle were brought to the surface, and those originally at the top and bottom were brought in contact at the middle. By this means, the whole of the bars were subjected to a very similar degree of violence. By a second trial of their deviating power, they were then separately tested, and, having been numbered, had their individual deviations registered. Two of the plates, however, being found considerably weaker than the rest, were rejected, and the subsequent experiments made without them. Thirty of them being then remagnetised up to their original energy, they were progressively combined on the trial-board, and the magnetic power of the mass was examined (taking the mean deviation produced by the two poles of each series) at every addition from one to thirty plates,—beginning, in the first experiment, with the strongest plates, and proceeding in order with those of less and less tenaciousness; and, in the second experiment, taking the plates in the reverse order. The annexed Table shews the twofold series of experiments, together with the ratio of the increase of power,—and from the diminishing accession of energy observed, the absolute loss, by the deteriorating influence of the augumented tension, is finally determined.

No. of Plates.	First Combination.		ReverseOrder.		Ratio of Increase of Power.			
	Deviation.	Tangent.	Deviation.	Tangent.	Mean of Tangents.	Gain of Power.	Power of Plates Separately.	Loss by Increase of Mass.
L	II.	III.	IV.	V.	VI.	VII.	VIII.	IX.
1	16·8° ′	289	16·35° ′	298	293	293	290*	—
2	29·3	555	27·48	527	541	248	580	39
3	38·20	791	36·16	734	762	221	870	108
4	45·6	1004	42·21	911	957	195	1160	203
5	51·0	1235	46·37	1058	1146	189	1450	304
6	54·18	1392	50·15	1202	1297	151	1740	443
7	57·23	1563	53·50	1368	1465	168	2030	565
8	59·12	1678	56·8	1490	1584	119	2320	736
9	60·52	1794	57·49	1589	1691	107	2610	919
10	61·56	1875	59·5	1670	1772	81	2900	1128
11	63·12	1980	60·3	1736	1858	86	3190	1332
12	64·4	2056	61·3	1808	1932	74	3480	1548
13	64·44	2119	61·49	1866	1992	60	3770	1778
14	65·23	2183	62·28	1918	2050	58	4060	2010
15	65·48	2225	63·4	1968	2096	46	4350	2254
16	66·10	2264	63·43	2025	2144	48	4640	2496
17	66·40	2318	64·10	2066	2192	48	4930	2738
18	66·58	2352	64·45	2120	2236	44	5220	2984
19	67·10	2375	65·10	2161	2268	32	5510	3242
20	67·19	2393	65·42	2215	2304	36	5800	3496
21	67·25	2404	66·0	2246	2325	21	6090	3765
22	67·33	2420	66·28	2296	2358	33	6380	4022
23	67·37	2428	66·48	2333	2380	22	6670	4290
24	67·54	2463	67·5	2365	2414	34	6960	4546
25	67·35	2424	67·21	2396	2410	− 4	7250	4840
26	67·26	2406	67·45	2444	2425	+ 15	7540	5115
27	67·31	2416	68·1	2477	2446	21	7830	5384
28	67·22	2398	68·16	2509	2453	7	8120	5667
29	67·17	2389	68·32	2543	2466	13	8410	5944
30	67·22	2398	68·54	2592	2495	29	8700	6205

* The average power of the different plates was about 16° 10′, tangent 290, from which average, column VIII. has been calculated— the loss by combination, column IX., being the difference of columns VI. and VIII.

The *advantage of combination* herein — in the yielding of additional magnetic power — becomes very obvious, when we compare the powers of the single bar magnet, (described at p. 101,) with an equal mass of these plates. One of the single bars weighing 6·236 lbs. was found, when fully magnetised, to acquire a deviating power of 54° 12′, of which the tangent is 1386; whilst fifteen plates, in close combination, of steel of similar quality, temper, and dimensions, being of nearly the same weight, (6·150 lbs.), exhibit a mean power of 64° 30′, the tangent being 2096.

After the completion of the series of experiments of the "Reverse Order" in this Table, the thirty plates were again separated into five parcels of six each as before (the six best plates being called series A, and the next in order of strength B, etc.), when their respective conditions, as shewn by their influence on the compass, were found to be as follow : --

Series.	Deviation at One Length Distance.			
	N. end.	S. end.	Mean.	Tangent.
A	46·4	46·14	46·9	1041
B	43·18	42·0	42·39	921
C	41·0	39·30	40·15	847
D	37·26	35·36	36·31	740
E	9·46	11·4	10·50	191

The sum of the tangents of these five series being 3740, whilst the highest tangent of the whole in one mass (thirty plates, col. v.) was only 2592—it was hence found that the deterioration by combination was not *altogether* permanent; but *partially recoverable on separation.* In this instance, the recovery of power as expressed by the difference of the tangents (3740— 2592 = 1148) was equal to almost one-half of the entire power when the thirty plates were combined in one mass.

But this does not express the whole of the suppressed power, capable of being recovered by the individual separation of the plates. For the determination of the total quantity of power, which might be considered as in a state of *elasticity*, it was necessary still to ascertain the actual state of the plates, after their full combination, *separately*, so as, by the sum of their respective tangents, to determine the entire amount of power recovered.

Proximately, and sufficiently near for the purpose herein contemplated, trial was made of the individual powers of two series of six plates —series A and C. The power of each series in combination was first determined, *immediately before* the examination of the plates separately — because the whole mass, having lain for some days together in contact, had sunk below the power exhibited in the Table. The state of each of the two series in combination, and with their plates tried separately, is exhibited in the Table following.

Series A.			Series C.		
No. of Plates.	Mean Deviation.	Tangent.	No. of Plates.	Mean Deviation.	Tangent.
6 Plates.	42·5	903	6 Plates.	40·45	862
No. I.	11·10	197	No. I.	8·17	146
II.	11·2	195	II.	10·8	179
III.	9·43	171	III.	8·40	152
IV.	10·32	186	IV.	9·20	164
V.	10·7	178	V.	10·7	178
VI.	6·48	119	VI.	7·25	130
	Sum -	1046		Sum -	949

From hence we find a still further and considerable recovery of the power which had been in a state of elasticity in each of the series—the gain in series A, by total separation, indicated by the difference of the tangents 903 and 1046, being betwixt one-sixth and one-seventh; and in series C, about one-tenth.

Another circumstance to be noted, in regard to the remarkable effects of combination on the weaker bars or plates, is the practical fact, that such effects result not so much from inferiority in capacity of these bars, as from the measure of violence to which they are subjected. For it was always found, as indeed theory would

have anticipated, that any change made in the
number of the plates combined, made a material
alteration in the comparative results; for the
plates which, in the more numerous combination, almost entirely lost their power, when
placed under less severity of tension or violence
in smaller masses, were capable of retaining a
very considerable proportion of their original
energy. Thus the whole series (omitting two),
as divided into five sets of six plates each,
commencing with the six strongest, marked A,
and gradually proceeding with the less tenacious
ones to set E, and then being remagnetised,
gave the following results : —

Set.	Mean Deviation of the two Poles.
A	53·25
B	50·45
C	50·56
D	49·45
E	48·15

So that the six plates of set E, which, when
in the mass of thirty, had their total power
reduced to about 9° 46′ = tangent 191, were
found capable of retaining, with tolerable permanency, a power of 48° 15′ = tangent 1120,
exhibiting, when in this more limited combination, a power nearly six times greater than
before !*

* No account is here taken, or in any other of the deductions, as to *proportional powers* of magnets, mentioned in this
Part of the work, of the "equation of deviations" described

Results, in all essential particulars analogous to those which we have been now describing, were obtained from experiments made with *bars* instead of plates. In one case, a series of ten bars of 13·8 inches in length, and each weighing rather more than a pound, was employed in a similar course of experiments to that constituting the leading subject of this Chapter; and, in another case, a combination of bars of two feet in length were tried in a similar way, but with no other essential difference in the phenomena being observed, except what was to be ascribed to inequalities in hardness, the effects of which *in combination* will have to be described more particularly in a subsequent chapter.

In the Diagram, No. 2, Plate iii., the respective powers of various combinations of magnetised cast-steel plates, of twenty-four inches length, as placed in contact, are shewn by the course and position of the lowest and thickest curve; whilst the degree of superiority of the power of such plates, over equal masses in a solid bar, or two bars together, is exhibited in their comparative heights above the general line of abscisses.

in Chapter iii. Part i., and represented to the eye in diagram, Plate ii., because the correction does not affect the *general results*, but only, in a slight degree, the numerical proportions.

RESULTS.

THE general Results deducible from the foregoing investigations, as to the laws or effects of
combination in magnetic bars or plates in contact—being applicable to the case of compound
magnets generally—may be now advantageously
collected, and, where needful, more particularly
illustrated.

1. That any *single* bar or plate is more powerful or energetic, *proportionally*, than two or more
corresponding and equal bars—that is, being
bars of like dimensions, quality of steel, temper
and mass.

This law is deduced from the invariable fact, that whatever may be the power of two or more bars combined
in one mass, the amount of their separate powers is
always greater. The proportion of *loss* by increase of
mass, in the large series of thirty plates, is shewn in
col. IX. of the table at page 103. Thus the power of
two plates (Nos. 1 and 2) in *combination*, is expressed by
the tangent 541 (col. VI.): the amount of their united
powers, *taken separately*, being 580 (col. VIII).

Hence, if a compass or variation needle were to be constructed in which the *mass* was of no consequence, or
which might be made indefinitely light, or requiring the
least possible momentum—then a single magnetic plate,

as thin and light as possible, would be the best. But as, for all ordinary uses in magnetical instruments, a given mass and momentum are absolutely necessary, for over-coming the resistance of the air and abiding its slightest motions: and as, when a compass card, collimater, or other substance or apparatus, is to be appended, a very considerable mass is generally requisite,—an important advantage is derivable from combination, as will forth-with appear.

The same law, though differing in quantity, applies to the comparison of single bars of various thicknesses or masses, being in all other respects alike—the thinnest bars being always more energetic, proportionally, than the thicker.

Thus, taking an example of the ratio of the power and the mass from the table at page 57, we find, in respect to the series T (tempered throughout), the tangent ex-pressive of the energy of the thinnest bar to be 191; whilst that of the next in the series, being double of its mass, was only 307, or little more than one-half more. And comparing the thinnest bar (v.) with the thickest (i.) of the same series, we find, that, whilst the bar i. was nearly nineteen times the weight of the bar v., its energy when fully magnetised, compared with that of the other, was only in the proportion of 621 to 191 (according to the tangents of their respective deviating effects on the compass), or but little more than 3 to 1.

2. That a *combination* of magnetic bars or plates is always more powerful than any *single bar* of equal weight to that of the whole mass combined, — the bars compared being made of the same steel, and alike in their mode and degree of tempering, form, and dimensions, as to length and breadth.

Though, as just stated, a single bar is more energetic in
proportion to its mass, than any number of equal and
corresponding bars,—yet a combination of bars is always
actually more powerful than the same mass in one bar,
as stated generally in Chap. i. Part i. and as more
satisfactorily appears by comparing the results of the
table at page 103, with the powers of different single
magnets—weight for weight. Thus, a pair of magnets,
of the same length and breadth as the plates under
examination, weighing 2·854 lbs. each, were found
capable only of deflecting the needle of the compass, at
two feet distance, in an angle of 35° 20',—but seven
plates, of the weight of 2·870 lbs, being very nearly the
same mass, produced a deviation of 53° 50' in one trial,
(col. iv.) and of 57° 23', in another (col. ii.), the mean
of the tangents indicating above double the power of
either of the single bars.—In like manner, though not
to the same extent of advantage, fifteen plates of the
compound series exhibited a superiority of power, over
a single bar of like nature and quality and weight, in
the relation of 2096 (col. vi.) the tangent of deviation
produced by the combined plates, and 1386, the tangent of the deviation produced by the single solid bar.
A similar result, though not to an equal proportion of
advantage, was obtained by a comparison of a series of
plates, equal, in mass, to either of a pair of bars, by
M. Gauss, in the Observatory of Greenwich—though
these, I have intimated, are very superior and powerful
for single bars.

Hence the corollary—that the directive power of any
single needle, of the ordinary form and kind used in
compasses, dipping needles, or variation needles, would
be considerably augmented if the same mass were
divided into, or combined out of, two or more plates of

the same length and breadth. And as there is no needle or bar in ordinary use in compasses, or dipping needles, or other similar instruments, but what might be constructed in a compound form, these instruments, by this construction, are capable of receiving a much higher directive power than they usually have—independent of the advantages to be derived from better qualites of steel, and improved modes of hardening.

Combinations, however, of divided plates, transversely, do not afford the same advantage, but results of a converse kind. For any plate, so divided, is found to lose power on the whole. Nor do I find that any material advantage is derivable from the division of plates into narrow slips, longitudinally, so far as my experiments have gone.

3. That the *absolute gain* of power in the combined mass, by each additional plate or bar, *progressively diminishes.*

The diminution in the *proportional gain* of power, by adding bar to bar, is such, that, in the series of plates referred to, the last twenty-six, out of thirty, did not double the power of the first six. In a set of twenty-four smaller plates of greater proportional thickness, the proportional thickness being as 2·5 to 1, the *last twenty-two*, which were added to the fasciculus, did not quite double the power of the *first two*. Analagous results were obtained on trial of thicker bars.

The ratio of power in combinations of the two-feet plates (principally referred to in this chapter), laid on each other and in sensible contact, as examined by successive additions, *in sets of six plates* each, exhibits, very nearly, the geometrical series of one-half, one-fourth, one-eighth, one-sixteenth, the directive power of the first set of six plates being equal to *one*.

An inspection of the "mean of tangents" (table, p. 103), shews how nearly this ratio prevails. For convenience of comparison, the selection of powers may be thus exhibited—the power of the first six plates being called one.

No. of Plates.	Mean Tangent of Deviation.	Differences of Sets of Six.	Proportion of Additional Power.
6	1297		
		} 635	$\frac{1}{2}$, nearly.
12	1932		
		} 304	$\frac{1}{4}$, nearly.
18	2236		
		} 178	$\frac{1}{7}$, nearly.
24	2414		
		} 81	$\frac{1}{16}$, nearly.
30	2495		

Hence, were the bars or plates perfectly equal in quality of steel, dimensions, and tenaciousness of the magnetic energy, there would, no doubt, be a continual accession of power (though gradually diminishing,) *ad infinitum;* but, it is obvious, from the foregoing considerations and facts, that (could the contact be made perfect) no increase, short of infinite, would bring the whole power to two, or double the amount of that exhibited in the set of the two-feet plates herein referred to, by the first six plates.

This result, however, can only be considered, in the present instance, as the type of the general law; as in plates of a different quality or temper, and in bars of greater thickness, various modifications of the ratio must be expected.

Thus, it was actually found, that in the case of the other set of twenty-four smaller, but proportionally thicker plates, the *ratio* of the power with the mass combined, (comparing the effect of additional plates in *pairs,*) was

different, though the result was in accordance with the
general law'. In this instance, considering the power
of the first pair of plates as one, the addition of
successive pairs presented, proximately, the following
fractional, and somewhat irregular, series:—

$$\tfrac{1}{4}, \tfrac{1}{8}, \tfrac{1}{11}, \tfrac{1}{14}, \tfrac{1}{16}, \tfrac{1}{19}, \tfrac{1}{21}, \tfrac{1}{26}, \tfrac{1}{28}, \tfrac{1}{30}.$$

And in the case of a series of *bars*, such as that of the
bars of 13·8 inches mentioned at page 108, where each
bar was of the weight of above a pound, the first bar
gave a power, by itself, represented by the tangent of
deviation 309; whilst the whole of the series, ten in
number, gave no higher a power than that represented
by the tangent 613, being scarcely double of that of
a single bar.

4. That continued additions to a powerful
combination of plates or bars, cease to be beneficial beyond a certain limited extent, because
of the impracticability of obtaining a large series
of such plates or bars perfectly alike. The
weaker bars of a powerful series (that is, such
as are inferior in quality of steel or of a lower
temper), not only not adding to the magnetic
energy of the whole mass, but sometimes, from
their own polarity being reversed, absolutely
injuring the total effect.

Column vii. of the table, at page 103, shews the gradually
diminishing accession of power by each plate respectively; and col. ii., in the first combination, shews that,
a maximum, in that instance, was obtained by twenty-
four plates, there being no increase, but an actual loss,
by the addition of the last six.

Similar effects were observed in almost every case where

a large number of plates or bars (not being perfectly hard) were combined in one series; so that a maximum of energy was generally obtained by a *portion* of the series, which maximum was found to be diminished, often materially, by the unfavourable action of the weaker plates or bars.

5. That some deterioration takes place in the *permanent* individual energy of all the bars or plates in combination, by every addition of power to the mass; the measure of deterioration in the separate bars varying with their difference of strength, or of their tenaciousness of the magnetic condition.

The term "strength" has been introduced in a former chapter (Part i. Chap. iv.) to express that quality of magnets by which the tendency to deteriorate, or return towards the neutral state, is resisted. Hence, the measure of the deterioration in different bars, exposed to the same degree of magnetic tension, is inversely as their strength. Strength and hardness, as a general law, seem to be accordant—increase of hardness yielding increase of strength. One anomaly, however, as to a particular case, will hereafter require to be considered.

This result, just above stated, is shewn by the condition of the several series of six plates after the combination (page 104), as well as by the condition of the individual plates tried separately: whilst the general agreement of different experiments for determining the degrees of strength in all the bars or plates of a given series, sufficiently establish the law.

As to the permanent deterioration of the plates tried separately, we find that the series A was reduced by being combined in the whole mass of thirty, from an average of 16° 10′, deflecting power, to 9° 54′; and series C, from an average of 16° 10′ to that of 9° 0′ (table at page 106).

In other cases, in which more powerful combinations were made, we find that some bars or plates, *in all respects to appearance alike*, lose their power entirely, whilst others suffer comparatively little by the combination.

6. That the relation of the deterioration of the magnetic energy to the quality of the plates, when combined in large masses, affords an easy and satisfactory process for *testing* the quality of each bar or plate so combined, in respect to its relative measure of tenacity or strength.

This method was employed, as already mentioned, for the separation of the two-feet tempered plates into the five different series comprising six plates each, numbered from A to E. Many other large bundles of plates were, in like manner, separated into their particular qualities, as to corresponding strength; similar results (as to all material differences of quality) being obtained on every repetition of the experiment. Particular examples of this mode of testing are given at page 40.

In the selection and trial of small bars or plates of an equal length, analogous results are obtained —with the advantage of being capable of ready comparison both with each other, and with a certain standard—by the use of the *test-bar* described at page 42.

Hence, we obtain important practical guidance, for a superior construction of compound magnets, of a two-fold character;—first, as to the inutility of making *large* combinations of bars on the ordinary plan of constructing compound magnets; and secondly, as to the

importance of removing all the bars which have their
polarity either reversed or neutralized by the combination, or which have their magnetism so reduced as to
add nothing essential to the total energy.
In the case of the ten bars mentioned at the close of the
illustrations of the result, No. 3, p. 114, two of the bars
out of ten were found to injure the total energy to an
extent of more than one-tenth, and so as to destroy the
effect of two others.

7. That whereas the *weaker* plates lose the
greater part of their power, or become absolutely
neutralized, or even have their poles reversed,
when subjected to the violence of large combinations—yet the same are capable of very considerable power and retentiveness of energy, if exposed
only to an inferior degree of violence in small
combinations.

The actual power retained in the five series of six plates
each, after the combination of the whole in one mass, is
shewn in the table at page 104; the state of each series,
as indicated by its deflecting power on the compass at
two feet distance, being as follow : —

A	B	C	D	E
46°·9	42°·39	40°·15	36°·31	10°·50

But when the same plates, being remagnetized, were combined in no greater mass than that of their own series of
six respectively—then their powers (p. 107) were these:

A	B	C	D	E
53°·25	50°·45	50°·56	49°·45	48°·15

Hence, whilst the weakest set of plates E, was reduced,
by the violence of the action of the whole series of
thirty, to a deflecting power of 10° 50', we find it capable

of retaining such a power, when only combined to the extent of the six plates of its own set, as to deflect the compass needle, at the same distance, to an angle of 48° 15′—being a power, in the proportion of the tangents, six times greater than before.

8. That whilst magnetic bars in combination, (being a state of violence or constraint when similar poles are coincident,) suffer a *permanent* deterioration of power, from that of their individual capacity for magnetism when separate— they also suffer a further *transient* deterioration, as if the magnetic energy were in a state of elasticity—which proportion of deterioration is recovered on the separation of the bars.

The extent to which this influence prevails has been already shewn in the account of the experiments connected with the tables at pages 104 and 106. The recovery of power by the separation of the whole mass of thirty plates into five series of six plates, was nearly one-half; and the further recovery of power by the trial of the plates separately about one-eighth more!

Hence, the weaker bars which, when taken out of the mass, are found to retain a considerable portion of energy, may, whilst in the mass, be altogether without power; or may even have a reverse polarity, so as to cause an absolute loss to the mass.

Thus, the last six plates of the series of thirty (table at page 103), being the weakest of the series, actually diminished the power of the whole (see col. III.), though each of them, when tried separately, after the combination was taken down, was found still to possess a deviating power, in the original direction, of from 2 to 4 degrees.

CHAPTER II.

1. When it was found that so much of the
aggregate power of the separate plates or bars,
in a series of magnets, was lost by the combination in one mass *in contact*—it was but natural
to anticipate that, an inferior deterioration would
take place by a combination of a less concentrate
description,—the plates or bars being kept *out
of contact* by substances, not susceptible of the
magnetic condition, interposed.

The effect of such separation, which I had
long ago determined in a limited degree, I now
proceeded to investigate by means of the series
of thirty tempered plates. For this purpose a
sufficient number of blocks of wood of different
thicknesses—constituting circular discs 1·5 inch
in diameter—was prepared,—each being perfo-
rated in the centre, so as with facility to combine
along with the steel plates ; whilst additional
wires, furnished with screw nuts, were also
provided, adapted by their lengths to the extent
of separation which might be required in the
several masses so combined.

The first and most elaborate investigation—
as to the effect of separation on the progressive
accumulation of power in the whole mass of
plates—was made with interposed discs of 0·14
inch (about one-seventh of an inch) in thickness.
As however, the series so separated—like the
frame saws in a saw-mill—spread out to the width
of several inches (and in subsequent experiments,
with thicker discs or blocks, sometimes to the
breadth of more than a foot)—it was found con-
ducive to accuracy to test their powers at *two*
lengths, or four feet distance, instead of one
length, as in the former experiments. At the
same time, for the advantage of comparison with
the results obtained at one length distance, occa-
sional trials—in the progress of combination—
were made of the action of the series on the compass at one length; which, with other comparative
experiments, by means of various combinations
and bars of the same length, afforded the means
of determining, experimentally, the precise ratio,
unequated, of the directive forces for distances
of one and two lengths. It may be convenient,
perhaps, to give this determination before proceeding with the investigation proposed.

The following table exhibits a selection of the
comparative deviations, at one and two lengths'
distance, of different combinations of two-feet
plates.

No. of Plates in Combination.	Deviation at one length.		Deviation at two lengths.		Ratio of two lengths to that of one length; one length being = 1000.
	Mean.	Tangent.	Mean.	Tangent.	
6	55·50	1473	15·32	278	189
6	55·20	1446	15·15	273	189
6	54·46	1416	14·55	266	188
12	67·30	2414	24·5	447	185
18	71·35	3003	29 40	570	190
24	74·22	3574	33·48	669	188
1	15·52	284	3·5	54	190
30*	68·30	2539	25·50	484	191
			Mean .	....	188·75

* The last combination (that of 30 plates) was in *contact*, all the others were separated by discs of wood.

Hence, the ratio of the directive powers of bar magnets, for one and two lengths distance from the compass (as thus determined for magnets of two-feet in length), appears to be as 1000 to 188·75. Theoretically, the proportions might seem to be a little different,—being inversely as N^2—R^2: n^2—r^2, N being the near pole of the magnet, and R the remote pole, at *one* length distance; and n the near, and r the remote pole, at *two* lengths distance. For this proportion, calling the power at one length 1000, would give $\frac{5}{27}$ for that at two lengths, or as 1000 to

185·2. The difference is in some small degree to be ascribed to the want of application of the equation of the tangents of deviation, described in Part i. Chap. iii., but mainly arises, I presume, from the circumstance of the *polar* or *focal* lengths of the bars being different from their measured lengths. For were the entire magnetic forces of the bars concentrated at the very extremities, the ratio, corrected for the equation of tangents, must have been according to calculation; but as the resultant of the magnetic forces, in either half of the bars, constitutes a point or *focus* of attraction *within* the extremity (which by other experiments I found to lie at about one-twelfth of the length of the bars from the ends), the comparison cannot exactly hold, unless the two distances were those of *focal* lengths, instead of actual lengths of the bars.

Thus provided with the means of comparing the deviations obtained at the different distances referred to, the results of a progressive combination of the plates, with interposed discs of ·14 inch, become available for the object contemplated. The deviating powers of the series, so combined, on the compass at two lengths or four-feet distance, were as follow: —

No. of Plates.	Mean Deviation	Tangent.	No. of Plates.	Mean Deviation	Tangent.
A 1	3·5	54	16	28·4	533
2	6·39	117	17	29·9	558
3	9·6	160	18	29·40	570
4	11·32	204	D 19	30·30	589
5	13·47	245	20	31·18	608
6	15·32	278	21	32·6	627
B 7	17·8	308	22	32·31	637
8	18·39	338	23	33·11	654
9	20·5	366	24	33·48	669
10	21·34	395	E 25	34·10	679
11	23·2	425	26	34·37	690
12	24·5	447	27	34·57	699
C 13	25·15	472	28	35·9	704
14	26·28	498	29	35·38	717
15	27·6	512	30	35·55	724
Closed into Contact,			30	25·50	484

Here then—without carrying out the comparisons with the table at page 103, of the progress
of increase over the former mode of combination
—the advantage of the separation becomes sufficiently obvious by the mere comparative results
of the entire series of thirty plates, as separated
by 0·14 inch blocks, and as closed into contact.
For looking merely at the foot of the above
table, we find the power of the series, when *separated* (as represented by the tangent), to be 724,
and, when in contact, only 484,—exhibiting a
gain in energy, by this very limited separation,
of about two-thirds of the original power.

2. The effect of separation at other, and
greater, distances, was now tried by means of the
same plates, and these combined in the same

order,—that is beginning with the series A, or strongest bars, and proceeding progressively with the series B C D, etc. The actual, as well as comparative, results, observed under different measures of separation, are the most compendiously exhibited in a tabular form.

Comparative powers of two-feet steel plates, magnetized to the limits of their capabilities, when separated by different-sized discs or blocks — the distance of the plates from the compass being two lengths or four feet, and the mean deviating power of one plate being 3° 4'.

No. of Plates.	In Contact.			Discs of 0·14 Inch.			Blocks of 0·28 Inch.		
	Mean Deviation.	Tangent.	—	Mean Deviation.	Tangent.	Gain over Cont^t.	Mean Deviation.	Tangent.	Gain over Cont^t.
3	8·32	150	—	9·6	160	10	•9·12	162	12
6	14·34	260	—	15·32	278	18	16·0	287	27
12	20·58	383	—	24·5	447	64	26·25	497	114
18	23·45	440	—	29·40	570	130	—	—	—
24	25·10	470	—	33·48	669	199	—	—	—
30	25·50	484	—	35·55	724	240	—	—	—

No. of Plates.	Blocks of 0·40 Inch.			Blocks of 0·48 Inch.			Blocks of 1·0 Inch.		
	Mean Deviation.	Tangent.	Gain over Cont^t.	Mean Deviation.	Tangent.	Gain over Cont^t.	Mean Deviation.	Tangent.	Gain over Cont^t.
3	9·16	163	13	9·18	164	14	9·40	170	20
6	16·28	296	36	16·54	304	44	17·44	320	60
12	27·38	524	141	28·39	546	163	30·51	597	214
18	—	—	—	37·48	776	336	—	—	—
24	—	—	—	43·45	952	482	—	—	—
30	—	—	—	47·41	1098	614	—	—	—

N.B.—The "gain over contact," is the difference of the tangents of each division, and the tangent of deviation when in contact. Thus, 150 (ool. iii.)—160 (col. vi.)=10 (col. vii).
These are not actual observations, but approximate numbers.

That a comparison of these various results
may be more readily made, and an accurate idea
conveyed, by mere inspection, of the proportion
of advantage gained by the system of combination—the relations of the several investigations
have been represented by the diagram No. 2, in
Plate III. For the power of the combined plates,
in the progressive augmentation of their number, is very well represented, after the manner
adopted in a former case, by a curve—of which
the abscisses represent the number of plates, or
mass of steel in the combination, and the ordinates the tangents of the angles of deviation for
the several masses. The diagram exhibits the
powers of various combinations; but those with
which we have at present to do, are the results
contained in the foregoing tables—the different
continuous curves, representing the powers of
the plates in their respective degrees of separation,—the lower and blacker curve (before
referred to) shewing the comparative power of
the same magnetized plates when combined in
contact.

Before I proceed to deduce the general results
from this mode of spaced combination, I may
mention the effect of *partial* contact and partial separation—such as the loss of power when
the plates, which had been put up in the usual
mode of spacing, were brought into contact at

one or both ends. For this investigation the thirty plates were combined in separate series of six plates each, in the manner before described, with 0·14 inch discs interposed. Their powers on the compass, at one length distance, being respectively ascertained, the effect was noted when the discs at the ends of the series were removed, and the plates pressed together at the ends, whilst the mass rested on the trial-board in the position in which its power was determined ; and, subsequently, the centre discs were also taken away, so as to bring the six plates into contact throughout. The results, at one length distance from the compass, were as follow :

Series.	Separated by Discs.		Closed at the ends.		Contact throughout.	
	Mean Devia-tion.	Tangent.	Mean Devia-tion.	Tangent.	Mean Devia-tion.	Tangent.
B	54·27	139·9	53·47	135·6	53·1	132·8
C	53 5	133·1	51·56	127·7	51·26	125·4
D	55·53	147·6	54·37	140·8	53·46	135·6
E	51·17	124·7	50·4	119·5	48·54	114·6
	Mean	136·3	....	131·1	....	127·3
	Loss per cent........			3·9	Loss per cent. 6·6	

Here the average *gain* by separation, and *loss* by partial or entire contact, are distinctly shewn. Even in this limited quantity of plates, amounting only to six in each series, we find the advantage of entire separation by means of thin discs (0·14 inch) over *contract at the ends*, was as 136·3 to 131·1, or as 100 to 96·2 ; and the advantage of entire separation over entire contact, was as 136·3 to 127·3, or as 100 to 93·4. In this case, the separated series is found to be only about one fourteenth part more energetic than the same series in contact; but when larger series are employed, the difference becomes much more considerable, so that thirty of these plates, separated by nearly half-inch spaces, attain a degree of magnetic energy more than *twice* that of which the same plates are susceptible when in contact.

RESULTS.

THE general Results derived from the investigations described, or referred to, in this chapter, were, in the main (except as to measure or quantity), similar to those yielded by the experiments on the combination of the same plates in contact. The following modifications, however, or additions, may be mentioned.

1. That the *quantity of gain* in power of any combination of plates or bars, over equal masses in a solid bar, is greater, where the plates or bars combined are prevented from coming into contact; and this augmentation of *gain* in power, *improves*, as the spaces betwixt the plates or bars are enlarged.

> This result, generally, was fully anticipated; but the proportion of effect due to different measures of space betwixt the plates could only be determined by actual experiment.
>
> Comparing the results of the table at page 124, we find the progressive gain, by the interposition of thicker discs or blocks, was such, that whilst a combination of six plates with spaces of 0·14 inch exhibited a power 6·9 per centum greater than that of the same in contact, —these six plates with spaces of one inch gave 23·1 per centum more power than when in contact. That a

series of thirty plates, with 0 14 inch discs, exhibited
a gain, in the whole, of 49·6 per cent., and with blocks
of 0·48 inch, obtained no less than 127 per cent. more
power, than when the same were in visible contact.

The Diagram, No. 2, Plate III., exhibits to the eye the
several relations of the power of the plates, as combined
in *contact*, or as spaced with interposed discs or blocks
of 0·14, 0·48, and 1·0 inch.

2. That a *larger number* of plates may be combined with advantage, and a greater measure of
directive energy obtained, by the separation of
the plates or bars—the proportion of increase of
attainable power, and the extent of advantageous
combination, having relation, with plates of the
same *temper*, to the measure of the spaces betwixt
them.

Hence it is evident, that the spaces might be so increased,
that the deteriorating influence of the plates on each
other would cease to be at all considerable,—and so
that, if additions continued to be made to the utmost
extent of *practicable* combination, each additional plate
of similar *strength*, would add a portion to the general
amount of power.

3. That the measure of *permanent deterioration*
in the magnetic energy of the several plates, is
respectively less, in proportion as the density of
the mass is diminished by separation.

4. That *weaker plates* can be combined to a
much greater extent, and with superior advan-

K

tage, when separated by discs or blocks, than
when the same are in contact.

5. That a *partial separation* — such as that
of the middle of the plates, the ends being in
contact — yields an additional energy to what
is obtained by entire contact, — the additional
energy, however, being by no means so considerable as when the separation is complete.

The proportion of advantage by separation throughout,
over partial separation in the middle only, in the case of
combinations of six plates, has already been mentioned,
as derived from the investigations in the Table at page
126; but that proportion becomes greater as the extent
of the combination increases.

For in the case of *entire* separation, we find, from the particulars in the Table at page 124, that whilst twelve plates,
separated by blocks of 0·48 inch, exhibited a power,
compared with the same in contact, of 546 to 383;
thirty plates so separated, compared with the same in
contact, obtained a power in the proportion of 1098 to
484. So that the advantage of this measure of separation was, — with three plates about 9·3 per cent. more
than the energy in contact; with six plates 15·8 per
cent.; with twelve plates, 42·5 per cent.; and with thirty
plates, 127 per cent. gain over the power in contact.

Hence a similar improvement, though inferior in degree,
was expected to be obtained by *partial* separation, over
contact, *with larger combinations* than those above described. In the case of thirty plates in combination, the
advantage of partial separation, over contact, as well as
the disadvantage of *any contact*, in comparison of entire

separation, is strikingly shewn by this one experimental fact:—that the thirty plates, which, in entire contact, had only a deviating power (at two lengths from the compass) of 25° 50' (tangent 484), previously exhibited a power of 31° 50' (tangent 621), when only *one* *end* of the plates was closed—being the end most remote from the compass;—whilst the same series, when fully separated by discs of 0·14 inch, had, in the first instance, before any contact was permitted, a power of 35° 55' (tangent 724). See Table at page 123.

In this experiment, the great disadvantage of closing the ends of the combination of plates, especially of a powerful combination, is remarkably proved by the comparison of the powers of the series when entirely separated, and of the same when closed at one end, and that the most remote from the compass,—the reduction of deflecting energy, in this case, being in the proportion of the tangents 724 and 621.

CHAPTER III.

WHEN these Investigations were originally made,
I was not aware that some of our early magneticians had recommended hard steel to be employed in the construction of artificial magnets,
—thus indicating that there were certain advantages, especially with relation to the *permanency*
of the power, to be derived from such a condition. Of the actual determination, however, of
the fact of any other advantage from absolute
hardness, than that of permanency, or of the
actual investigation of the relative effects of
various degrees of hardness,—I have not met
with any record. A certain degree of hardness,
for compound magnets, had been mentioned as
desirable; such as that produced by tempering
the steel at the heat of "cherry-red" (a moderate
degree of hardness only); but this recommendation rather went *against* the supposition of
greater degrees of hardness than that being
generally advantageous.

The field of investigation, therefore, entered

upon in this chapter, may still, I apprehend, be considered as new,—so far, at least, as regards the determination of the actual and comparative influence of the various degrees of hardness or qualities of temper, by which the more important properties of magnets may be influenced.

Much, indeed, of this portion of our subject, has been partially involved in the investigations in Part I.—" on the ratio of the power of magnetised steel bars, as modified by difference of tempering;" and many of the general results might, by analogy, have been anticipated. But still, the investigations on the actual effects of tempering and hardness, on *combinations* of steel plates and bars, will not, I hope, be found unimportant, especially as affording additional information of much practical consideration in the construction and improvement of magnetical instruments.

A summary of, or selection from, the investigations made on this subject—with bars or plates in several distinct conditions, as to temper —will afford all the information, as yet obtained, which has appeared to me to be of sufficient importance to be recorded.

1. *Of the relative powers of combinations of magnetised steel plates, as affected by* changes *in the temper of* the same *series of plates.*

Captain Kater, in his investigations for his Bakerian lecture, published in the Philosophical Transactions, arrived at the conclusion, that increased energy was given to the small magnets designed for compass needles, by *reducing* the hardness in the middle of the needles after their having been tempered at a blue throughout.

As, however, the investigations of this scientific experimentalist had been chiefly made with small masses of steel, and, generally, with thin plates, it by no means followed that this was a general law,—as we shall hereafter find it did not in the case of other results obtained with like masses. On the contrary, it will be abundantly proved, that the *principles* which affect the proportionate powers of small and thin magnets, of unequal degrees of hardness or difference of quality, will not apply to a comparison of the powers of large masses, under the like conditions.

In order to determine what might be the effect, of reducing the hardness of a combination of tempered steel plates, on their magnetic capabilities, I made trial of different sets, some being two feet in length, some 8·7 inches, and others six inches;—the plates of each set being of corresponding size, quality, and temper.

The first set, subjected to investigation, consisted of plates of ordinary steel, cut out of a parcel obtained at the shop of an ironmonger and tool-maker in London, 8·7 inches in length, ·65 inch in breadth, about ·035 inch in thickness, and weighing, on an average, 370 grains. These were 24 in number, consisting of a selection, out of the original series, of the weaker plates, or those of an inferior degree of retentiveness. Their degree of hardness was, I considered, that of a low spring temper, pretty regular throughout.

The whole series, being strongly magnetised, was first laid together in one fasciculus, and their order of tenacity, for magnetism, individually determined. Each plate, having been previously marked with a separate number, had its several capabilities registered.

Two sets of experiments were then undertaken with this series, in their *original condition* as to temper, arranged in their order of retentiveness or strength, for the determination of the progressive accumulation of power; first, commencing with the strongest plates, and next, after their being remagnetised, commencing with the weakest, or in the reverse order of the series.

These experiments being completed, the whole of the plates were then *reduced* in the middle to the extent of about six inches, by laying them in succession across an iron bar, heated to redness, till the surface above became of a deep purple colour, and then letting them cool slowly,—

leaving nearly one and a half inches of each end
in the original state. In this condition they
were re-magnetised, and their powers, in progressive combination, again determined.

Twelve of the series, from the weakest side
of the entire fásciculus of plates, were then
heated *throughout their length* in the same manner
as before, and their powers also examined.

The whole of the results are given in the
following table. The first division exhibits
the actual and comparative powers of the whole
series, commencing with the strongest plates, in
two different conditions,—that of their original
temper, and that of their supposed *reduced* state,
after being heated in the middle. The second
division shews the powers of the series in a
reverse order, commencing with the weakest
plates, in their original state as to temper—with
the comparative powers of a part of the same
series when reduced *throughout* their length, by
heating as before.

TABLE exhibiting the Comparative Powers, in combination, of the Series of smaller Plates, 8·7 inches in length, in three several states of temper.

No. and Order of Plates	Power in original tempered state.		Power when reduced in the middle.		Order of Plates	Power in the original temper.		Power when reduced throughout.	
	Mean Deviat.	Tangent	Mean Deviat.	Tangent		Mean Deviat.	Tangent	Mean Deviat.	Tangent
1	9·42	171	10·56	193	24	8·52	156	8·58	158
2	15·55	285	17·54	323	23	14·22	256	14·29	258
3	18·21	332	20·46	379	22	16·49	302	17·9	309
4	19·22	351	21·55	402	21	18·0	325	18·33	336
5	21·7	366	23·34	436	20	19·26	353	19·54	362
6	21·38	397	23·42	439	19	20 13	368	20·54	382
7	22·25	413	24·31	456	18	20·49	380	21·34	395
8	22·44	419	25·5	468	17	21·15	389	22·6	406
9	23·33	436	25·32	478	16	21·58	403	22·40	418
10	23·30	435	25·47	483	15	22·27	413	23 31	435
11	23·55	443	26·17	494	14	22·56	423	24·4	447
12	24·25	454	27·2	510	13	23·27	434	24·48	462
13	24·45	461	27·43	525	12	23·50	442		
14	25·18	473	27·53	529	11	24·26	454		
15	25·38	480	28·15	537	10	24·42	460		
16	26·12	492	28·44	548	9	25·22	474		
17	26·30	499	29·0	555	8	25·43	482		
18	26·43	503	29·25	564	7	26·7	490		
19	27·13	514	29·52	574	6	26·40	502		
20	27·35	522	30·16	584	5	27·13	514		
21	27·42	525	30·23	586	4	27·38	524		
22	28·4	533	30·44	595	3	27·56	530		
23	28·28	542	31·15	606	2	28·18	538		
24	28·41	547	31·16	607	1	28·47	549		

The results of these several effects of heating, and (as it is assumed) reducing in temper, were not such as had been expected. For, from all foregoing experiments, I had been led to conclude, that, whatever might have been the effect of the reducing of the temper in single thin plates, or small masses, the effect in *combination* would have been to diminish the energy of the plates,—whereas a considerable gain of

power was yielded by the heating of the middle
of the plates, and some increase beyond the
original power, though not to the same extent,
was produced by a corresponding heating of the
plates from end to end.

This discrepancy with the results previously
obtained, having occasioned some embarrassment and misgiving as to the correctness of the
principle I had considered certain, — namely,
as to the advantage of an equable hardening
throughout for bearing the tension of powerful
combinations of magnetised bars or plates,—
I sought to verify the newly obtained result, or
to determine the cause of the *apparent* discrepancy, if not real, by a trial of the same process with a set of larger, and, proportionally,
thinner plates. For this purpose I selected a
set of six plates of two feet in length (that of
series E referred to at page 104), and reduced
the temper in the middle of each plate, leaving
about three inches in length, at each end, in
its original state.   This was done in the same
manner and proportion as in the first alteration
made in the temper of the smaller plates, just
spoken of, and by which treatment the most
considerable increase of power in these plates
had been yielded.

The change which took place in the power of
each plate by this alteration of temper, as ascer-

tained by the deviation produced on the compass
at the distance of two lengths, or four feet, was
not the same as that indicated by the former
experiment, as here distinctly shewn :—

No. of the Plate.	Mean Power of Plates separately.	
	In their original state.	After being reduced in the middle.
1	3·17	3·15
2	3·0	2·53
3	3·1	2·55
4	3·8	2·53
5	3·0	2·59
6	3·8	3·7
Mean . .	3·5⅔	3·0⅓

This result, it will be perceived, appears to
be the very reverse of that obtained by a similar
mode of treatment on the set of 8·7 inch plates.
For in that case, a partial heating of the plates
in the middle, until they assumed a purple
colour, augmented the capacity for magnetism
of the whole series, when in combination, in the
ratio of 607 to 547, or about one-tenth gain ;
whilst a similar heating of a dozen of the same
plates from end to end, gave an increase of
power, in the combination, over that of their

original temper, in the ratio of 462 to 434,
being a little more than one-fifteenth gain. But
in this subsequent experiment with the set
of six two-feet plates, a *loss* of power was
sustained by every one of the plates, amounting,
on an average, to about one thirty-sixth part of
the original energy. A trifling loss was likewise
observed to have taken place in the tenaciousness of the plates,—the power of the six plates
in combination and contact, after the reduction
of the temper, being less than that of the same
plates in their original condition.

In order, however, to investigate still further
the cause of the discrepancy of the results with
the smaller plates, I ordered another set of two-
feet plates of best cast-steel, comprising two descriptions, as to their kind of tempering. They
corresponded, as to dimensions, with the series
of thirty plates, so much already treated of, except
as to thickness,—these new ones being of the
average thickness of ·057 inch, and of the weight
of 545 pound,— the former averaging ·042 inch
in thickness, and ·410 pound weight. This new
set of plates consisted of twenty-four in number,
of which sixteen were tempered equably throughout, a little *harder* than the former, and the other
eight were tempered in a similar way at the
ends, being *softer* in the middle. The average

weight of the plates of each kind was very nearly alike.

All these plates being magnetised fully, by at least four strokes (two on each side) by process KS, were examined as to their deviating powers severally, at the distance of two lengths, or four feet, from the compass; each kind was then tried in combination, separated by discs of ·14 inch (the sixteen harder plates in two fasciculi of eight), and so their various powers were determined. After some other experiments, hereafter mentioned, the twenty-four plates were placed in one mass in *contact*, with similar poles coincident, not in their original order, but with the plates of each series of eight regularly alternating, so that the whole were thoroughly changed and intermixed. The power of each plate, after being subjected to this state of violence, was finally examined, by which the relative strength or tenacity of all the plates for the magnetic condition was found, and a mean of comparison, betwixt the two kinds of plates, as to tempering, obtained. The last column in each division of the annexed table, shews the individual power of each plate after the final combination,—the other columns their powers when first magnetised (determined separately), as also their powers in combination when separated by the discs.

	EQUABLY TEMPERED PLATES.							PLATES SOFTER IN THE MIDDLE.			
No. of the Plate.	Separate mean deviating Power.	Mean Power in combination.	Separate Power after contact of the 24 Plates.	No. of Plate.	Separate deviating Power.	Mean Power in combination.	Separate Power after contact.	No. of Plate.	Separate deviating Power.	Mean Power in combination.	Separate Power after contact of the 24 Plates.
1	3·45	3·45	2·15	9	3·34	3·34	1·55	17	3·38	3·38	1·25
2	3·9	6·35	1·50	10	3·8	6·9	1·15	18	3·17	6·28	0·16
3	3·33	9·25	0·50	11	3·40	9·10	2·15	19	3·31	8·58	1·42
4	4·25	12·15	3·4	12	3·42	11·43	2·20	20	3·26	11·19	0·47
5	3·37	14·27	2·5	13	4·7	14·17	2·47	21	3·18	13·8	1·38
6	3·31	16·10	1·16	14	3·33	16·15	1·17	22	3·30	14·55	0·45
7	4·40	18·25	3·10	15	3·11	17·35	0·47	23	3·24	16·20	1·30
8	4·15	20·15	2·55	16	3·11	18·58	1·15	24	3·27	17·55	1·0
Mean	3·52	.	2·11	.	3·31	.	1·44	.	3·26	.	1·8

Here we find a very decided difference in the
two kinds of plates (both as to individual capa-

city for magnetism, and as to power in combination), in *favour* of the plates tempered or
moderately hardened throughout. The separate
powers of the softer plates, indeed, were very
nearly equal to those of the second series; but
after being subjected to the violence of contact
in the whole mass, their inferiority became very
apparent. The mean power of the softer plates,
separately, compared with that of the second
series of harder plates—was as the tangent of
the angle of deviation, 3° 26′, to tangent of
3° 31′; but after combination, as tangent of
1° 8′ to tangent of 1° 44′.

The inequality in the *strength* of the different
series, however, became more apparent, when
they were *tested* by being all combined in one
fasciculus, with similar poles together. The
resulting powers of the several series then tried
separately—that is in sets of eight plates each—
were found to be very dissimilar, the first set
producing a deviation in the needle of the compass of 14° 7′, the second of 12° 5′, and the third
(or that soft in the middle) of 8° 5′.*

---

* It is proper to be noted, that an undue application of
these results, with reference to the particular question under
consideration, be not made, that the softer condition of the
middle of some of these plates was, probably, not produced by
a second heating, or *reducing*, as in the former cases considered
in this section.

Notwithstanding the accordance of all these latter experiments, made with two-feet plates, with the results of investigations described in the former chapters,—the difference in the case of the smaller plates yet remained unexplained. One or other of these conclusions, therefore, seemed inevitable,—either that the heating of moderately tempered bars until a change of colour appears, does not necessarily *reduce* the temper, or degree of hardness; or that such reducing of temper, in *certain cases* (the exact conditions of which are not yet determined), instead of diminishing, serves, as Captain Kater had concluded, to augment the capacity of steel for magnetism.

Not being able to ascertain, whether the portions that had been discoloured by being heated were, or were not, reduced in temper,— I considered that I might determine satisfactorily, by another course of procedure, whether the reduction, if any, was at all considerable.

In order to the determination of this fact, I took a portion of the 8·7 inch plates (nine in number), which had been, apparently, the most reduced. These I put together into an ordinary kitchen fire, until heated to the apparent degree of the bright cinders amid which they were placed. On being removed, they were laid on the hearthstone to cool.

The magnetic condition was now found to have undergone a very great change. Magnetised to the utmost, their deviating power, separately, was, on an average, reduced from about 9° 17′ (the power in the original temper) to about 6° 55′; whilst their power in combination was reduced very nearly one-half! Or taking the proportions conversely, in respect to the superiority of one condition over another, we find, that although, in the former experiments, nine of the plates in combination which had been moderately heated in the middle, obtained an accession to their original power of about one-tenth; yet, nine of the same plates, as shewn by this latter experiment, when in their original condition of temper, had a greater power than the same had when heated to redness in the kitchen fire, in the proportion of 418 to 225, or nearly nine tenths more

Hence it becomes obvious, that, whatever may have been the change produced in the condition of the plates, by the former mode of heating on a bar of iron, the *quantity* of that reduction must have been very small.

The actual effect of this small measure of heating, on the arrangement of the particles of the steel in that particular state of elastic temper, I was disposed, at this point of the investigation, to imagine, might perhaps be only just sufficient

L

to disturb, or raise, as it were, the crystalline structure of the ferruginous elements, which, in the case of the larger and thinner plates, subjected to the same treatment, had been not only disturbed, but partially changed or *reduced!*

Another view of the difficulty, however, now occurred to me, by which, what had hitherto proved so embarrassing, seemed likely not only to be resolved, but the principles elsewhere deduced, established and confirmed.

It has before been shewn, from the experiments described in chap. v. Part i., and by the principles exhibited in chap. viii., that the energy which a magnet, unsustained by a keeper or conductor, is able to receive and *retain*, is dependent on the measure of concurrent action of the two qualities of capacity for magnetism, and of tenaciousness or strength. The first quality is possessed in the *greatest degree* by *iron*, especially by iron of the finest and softest descriptions; and next by *steel*, of the finest kinds and in the softest state. The second quality, which is scarcely possessed at all by iron, and in a very moderate degree by soft steel, belongs most eminently, in respect of steel, to that of the greatest hardness; and occurs, intermediately, in a certain relation to the degree of hardness. Hence these two qualities have a varying influence on the power of a magnet, according as

the mass is enlarged or the hardness increased;
so that in small masses and very thin plates,
capacity gives the pre-eminence, and in large
masses or thick bars, *tenaciousness,*—inasmuch
as in these, where the magnetic tension or
violence is great, the power of sustaining the
tension is of the first importance.

We find, therefore, as has been before shewn,
that there are particular limits within which
each quality, as it affects merely the primary
energy immediately after being magnetised, has
its pre-eminence.   If the tension of the magnet,
or combination of magnets, be not beyond a
certain limit, the moderate tempering will outvie
that of greater hardness; if beyond that limit,
the increased hardness will be superior.

Now the difficulty under consideration might,
it appeared probable, be resolvable into this
particular case—viz. a case in which there being
no great tension or violence of combination,
the gain of capacity, by the slight and partial
reduction of the hardness, had not been overcome by the effect of the diminished tenaciousness. For if this were the fact, then the apparent
difficulty would resolve itself into actual conformity with the previously determined principles on which the energy of magnets of different
masses and degrees of hardness is dependent.

And that this was really the fact, in the special

case contemplated, this consideration rendered probable, that, although the number of plates combined was so considerable (twenty-four), yet their united energy was but of very moderate amount,—so moderate, indeed, that (as the inspection of the table at page 137 will shew) the total effect of the whole series was but little more than *three times* that of a single plate!

The fact, however, promised to be determined by another mode of investigation, in which the inquiry should be,—whether, in cases where an augmentation of energy was yielded by heating or reducing the plates in the middle, that reduction of temper was not accompanied by a diminution in tenaciousness or strength? If so it should prove, then would the present case come strictly within the principles before laid down.

In order to the determination of this inquiry, I selected five plates of tempered steel, of the uniform length of six inches, of the breadth of three-eighths to one-half inch, and of different thicknesses. They were all supposed to be of a spring temper throughout. The maximum power of each of these plates was first determined, and the reduction, after application to the test-bar (H), ascertained on the trial-board.

They were then placed on a piece of iron, heated to redness, till the colour near the middle partially changed to a blue, when they were plunged into cold water; the maximum power of each, with the effect of the test-bar on it, was again determined. Finally, the same plates were heated again, to a little further extension of the discoloration, and placed on the hearth-stone to cool;—the powers, as in the other cases, before and after testing, being once more determined on the trial-board. The whole of the results are comprised in the annexed table:

No. of the Plate.	Weight in Grains	Original tempered state. Temper T.		Reduced in the middle and quenched.		Reduced again and cooled gradually.	
		Maximum Power.	After the Test Bar.	Maximum Power.	After the Test Bar.	Maximum Power.	After the Test Bar.
I.	II.	III.	IV.	V.	VI.	VII.	VIII.
I.	634	19·8	5·24	18·41	5·7		
II.	246	13·54	5·15	13·45	4·46	13·47	4·29
III.	162	10·39	4·29	10·45	4·10	10·44	3·55
IV.	147	10·8	— 1·0	10·10	— 1·4	9·56	— 1·13
V.	120	10·0	+0·42	10·10	+ 0·7		

These results completely verify, I think, the views just suggested, by which an apparent difficulty becomes not only accordant with, but confirmatory of, the general laws. For here, it will be perceived, by comparing columns III. and V., that, in three instances out of five (numbers III. IV. and V.), the reduction in the temper of the plates yielded a small augmentation of the magnetic energy—in accordance with the fact in the particular case inquired into. But this augmentation, it is observable, belongs *to capacity only*, and not to strength or tenaciousness. For the effect of the test-bar (constituting a more severe trial of the "strength" for retaining magnetism than that yielded by the combination of the 8·7 inch plates in one fasciculus), was uniformly to shew a deterioration in tenaciousness. Comparing columns vi. and viii. with column iv. (the latter shewing the power of the respective plates under the action of the test-bar in their original condition, as to temper), we find an undeviating verification of the previously held law,—viz. that, whilst a reduction of hardness may, within certain limits as to mass, yield increase of energy, it is obtained at the expense of a measure of tenaciousness. The gain, therefore, in the combination of the 8·7-inch plates, by a slight reduction of their temper in the middle, appears to be due simply to the increase of capacity for magnetism yielded to the plates by this change,—a gain, however, which, it now appears evident, would have disappeared, had the plates been subjected to a sufficiently high degree of violence.

That this is no arbitrary assumption, we may assure ourselves by comparing the *proportional* dvantage in the reduced temper (table, page

137) at the beginning and end of the series. In the first instance the increase of power in the plates *separately*, by the partial heating in the middle, was about one-fifth of that of the plates in their original and equable temper; but the *proportionate* gain, when the twenty-four were combined, had reduced to about one-tenth. So that, had the combination been carried sufficiently far, the set of plates which had been improved by being partially heated, would, in all probability, have been ultimately found inferior to those in the original state.

Having projected the powers of this series of plates in their different alterations of temper, after the manner of the diagrams already given, I found this result, apparently, inevitable;—for the production of the curves, so far as obtained, in a regular form, caused them to intersect at the place of about the thirty-second plate, and to yield, ultimately, (in exact accordance with theory) an unequivocal indication of superiority of power to the harder or original condition of the plates. [See Diagram, No. 4.]

Guided by this view of the once perplexing difficulty, the explanation of the differences in the results of the other experiments connected with the same inquiry, becomes, for the most part, abundantly simple. For it would hence appear, that the *loss* of power in the thin two-feet plates, by their being reduced in the middle

(table, page 139), is to be attributed to the over-
reduction of temper—of the effect of which, in
an extreme case, we have a striking example in
the thorough softening of the 8·7-inch plates,
described at page 145. In the case of the larger
series of two-feet plates, in which the powers of
those equally tempered, and those " softer in
the middle," are compared (table, page 142), we
infer a like condition, in explanation of the fact
of a loss of power in the plates of reduced
temper, viz., that they were softer in the middle
than was consistent with the modification of
structure resulting in gain of capacity.

In regard to the particular instances given in
the last table, in which the reduction of the
temper of two of the plates, *though very slight*,
was attended with a loss, instead of a gain, in
the resulting energy, other circumstances, relating to the quality of the steel, degree of hard
ness, or the original *mode* of tempering, might
perhaps, if fully known, have explained the
difference in the results.   But these inequalities
are of no consequence in the more immediate
inquiry; our object being to shew, as I trust
has been accomplished, that the *apparent discrepancy* of the results obtained by slightly
reducing the temper of the series of 8·7-inch
plates—as to their power in combination—does
not militate against the laws previously determined, nor does it invalidate the principle of

the mode of *testing* for ascertaining the degree
of hardness.

At all events, the embarrassment occasioned
by this particular case, in its bearing upon certain principles deduced from the former investigations, has been fairly and candidly stated, and
the series of experiments connected with it given
in detail. The scientific reader, therefore, will
be able to form his own judgment on the explanation just suggested—whilst the magnetician
may be induced, perhaps, to investigate the case
more fully, so as to come to a still more decided
conclusion in respect to the exact conditions on
which all the apparent differences depend.

2. *Of the magnetical powers, separately, and
in combination, of steel plates variously tempered,
both in extent of surface and in degree of hardness.*

The object of this investigation was to determine, not only the effects of various differences
in the degree of hardness, in similar plates, and
of inequalities of temper in the same plates,
upon their respective magnetic capacities and
tenaciousness; but also the relative effects of
such differences of temper on *combinations* of
plates similarly prepared, as compared with the
effects on *single* or separate plates.

Well calculated for the proposed inquiries was
a series of steel plates, of the nature of busks
used in the stays of females, tempered, in sets,

in a considerable variety of ways, which had been considerately prepared for me by Mr. Davenport of Sheffield,—a manufacturer whom I had employed for the construction of a considerable part of the bars and plates used in the foregoing investigations.

The whole series comprised twenty-one plates of steel of similar dimensions, very nearly,—the length of each being 14·4 inches; the breadth 1·5 inch; the thickness about ·025 inch. They were correspondingly tempered in sets of *three* plates, and comprised the following varieties, in the description of which the average weight of the plates of each set is specified:

No.		Average Weight. lb.
I.	The steel in the three plates of this set was described as being in the " raw state," as it comes from the rolling-mill, *quite soft* -	·140
II.	The same kind of steel, *hardened throughout,* as hard as glass - - - - -	·138
III.	Similar busks *tempered,* spring temper, in *the middle,* the ends remaining quite hard -	·133
IV.	Similar busks, *tempered at one end,* the rest quite hard - - - - - -	·137
V.	Similar busks, *tempered at both* ends, the middle quite hard - - - -	·133
VI.	Similar busks, tempered throughout, and quite elastic - - - - - - -	·139

[All the above were black and unpolished.]

VII.	Three busks, quite finished, tempered, glazed, perfectly elastic, select specimens - -	·132

With these busks—being all magnetised carefully by process KS—experiments, separately and variously combined, of which the following table contains the results, were made:

TABLE of Comparative Powers of Seven Sets of Magnetized Steel Busks, tempered in different ways, and subjected to different extents of combination.

[Deviations at Two Lengths from the Compass.]

Sets of Three Busks.	Mean Power of each Busk. Tangent.	Tangent.	Sum of Tangents of the 3 Busks.	Three Busks in combination, [Mean of two experiments.]			After the whole twenty-one had been in combination; two experiments.			State of separate Busks after combination.		
				Deviating Power.	Tangent.	Loss per Cent.	Deviating Power of the set.	Tangent.	Loss per cent. in the 3.	Mean Power.	Tangent	Loss per Cent.
I.	II.	III.	IV.	V.	VI.	VII.	VIII.	IX.	X.	XI.	XII.	XIII.
	° '			° '			° '			° '		
I.	5·87	98·3	295	11·25	202	33·5	−2·32	−44	113·0	−1·0	−17·5	118·0
I.	4·7	72·0	216	11·16	199	8·0	+10·40	+188	13·0	+3·43	+65·0	9·7
II.	4·44	82·8	248	12·59	231	6·8	9·36	169	31·8	3·26	60·0	27·5
III.	4·35	80·2	241	11·23	201	16·2	8·47	155	35·6	3·6	54·2	32·4
IV.	4·40	81·6	244	10·43	189	22·6	6·49	120	50·8	2·28	43·1	47·2
V.	6·39	116·6	350	15·6	270	22·8	2·30	44	87·4	1·21	23·6	79·8
VI.	6·15	109·5	329	14·33	260	21·0	8·55	157	52·3	3·29	60·9	44·4

The relative measure of retentiveness, or strength, in the different combinations, especially under that of the greatest violence when the whole series was laid together, we have here very strikingly and satisfactorily exhibited. For we have here proved, what theory suggested, that whereas the perfectly hard plates received the weakest power, in their separate condition after being first magnetised, yet these were found to be the most energetic after the whole series had been placed in a mass.

The comparative capacity for magnetism, separately, of the different classes of busks, is shewn by the angle of deviation, column II., or by the numerical tangent, col. III., and the sum of the powers of the three busks of each set, by the tangent in col. IV. Comparing col. IV. with col. VI.—the tangent of the deviation produced by the three busks of each set, laid upon each other and pressed together by a weight of about a pound—we obtain the proportional loss by this measure of combination, which is represented in col. VII. And comparing col. IV. with col. IX.—the tangent of the angle of deviation produced by each set in combination, after they had been laid in one mass, and the several busks of each set dispersed in an alternating series in the mass, and the whole compressed together by tying—we find the proportional loss produced by

this exposure to the greater violence, represented
in the ratio per centum in col. x. Comparing,
finally, col. iii. with col. xii.—the tangent of
the average deviation of the three busks of each
set, tried separately, after the combination of the
whole mass—we perceive the proportional loss
in the busks, severally, by this combination,
represented in col. xiii. at the rate per centum.

The influence of tempering, or the reduction of
hardness, in producing a diminution of *strength*
in the busks—was evinced in the case of the
series No. iv., tempered only at one end, in a
manner much more strikingly than appears in
the table. When first magnetised and tried
separately, the average power with the *hard*
extremity presented to the compass, was 4° 35′,
and with the *tempered* extremity the same: combined in their own series of three busks, the
hard end exhibited a deviating power on the
compass of 11° 38′—the tempered end of 11° 6′.
After combination in the whole mass, the three
busks together produced a deviation of 10° 0′ with
the hard ends towards the compass, and of 7° 53′
with the tempered ends towards the compass, —
the average power of the three busks separately
now being—hard end towards the compass 3° 45′
—tempered end 2° 27′.

In series No. i.—in which the busks were
perfectly soft—the magnetism was not merely

neutralised by the effect of the violence of the
mass, but the poles reversed. And in series
No. vi.—tempered throughout, the power was
almost destroyed. The great difference betwixt
the effect of the violence on sets No. vi. and
No. vii., is only to be ascribed to temper, which,
though *supposed* by the manufacturér to be the
same, was, no doubt, very dissimilar.

No account has been taken in these, or in any
of the foregoing, experiments, with similar plates,
of the little differences of weight in the several
plates,—because of the small corrections which
might be requisite for reducing the observations
to uniformity of mass in each plate, being much
less than the differences arising from want of
correspondence in the nature of the contact, in
the process of magnetising the plates, or from
varieties in the actual condition of corresponding
series, as to temper, or density of the substance.

The total energy of the mass of twenty-one
plates in combination, when bound together by
a string, was found, in one experiment, to be
equivalent to a deviating influence of 26° 53′;
and, in another, after the whole of the busks
had been remagnetised, of 26° 50′   But had
set No. i. been withdrawn, which had abstracted
from the total power, and also No. vi., the effect
of which was doubtful—the deviating power of
the remaining fifteen harder busks, would, without doubt, have proved much greater.

3. *Of the magnetic powers of combinations of*
very hard *plates and bars.*

Finding, from the various investigations
on the hardening of steel, that the condition
of hardness was so advantageous to magnetic
energy when tried both on single bars of
considerable thickness, and on plates combined
into moderate masses—it became important to
determine—what positive measure of advantage
the *greatest hardness* might yield, when the steel
so prepared should be subjected to the violence
of still more powerful magnetic combinations?

(1). *As to hard plates.*—In the trial of the
busks, employed in the experiments detailed in
the preceding section, it was found, that, although the capacity of the hard busks (No. II.)
for the magnetic condition, was the lowest of the
series when taken separately, yet when compared
with others as combined only in sets of three,
their united powers had gained an equality,
nearly, with several of the other sets. When it
was also found, that, subjected to the violence of
a combination of all the different sets together,
the hardest plates had advanced beyond the
power of any of the others;—it became quite
obvious, that a much greater *ultimate* power might
be obtained *in combination*, by the employment
of thoroughly hard plates, than by any of the

combinations of tempered plates hitherto employed.

For the determination of the real effect of such
combination, as to its comparative advantage,
a large quantity of busks, thoroughly and similarly hardened, was ordered of the manufacturer.
These consisted, in the first instance, of 72 similar
plates—but the number was afterwards increased
to 192. The common or average dimensions of
the whole, were, 15 inches in length, 1·5 inches
in breadth, and ·025 inch in thickness; and
the average weight was ·1534 lb. or 1074 grains.
They were all in a rough state, just as the plates
came out of the bath in which they were hardened.

The separate action of these plates on the
compass, at two lengths' distance, was such as
to produce a mean deviation of 3° 21′.

The following table exhibits a twofold series
of experiments: first, of the powers of 72 plates;
and, secondly, of those of 192 plates. In the
first series, the powers in combination of 12
plates are given in a regular series, and then
the effects of further additions of two, six, or
twelve plates, to the fasciculus, up to the amount
of 72. In the second series, the powers are
given in successive additions of 24 plates after
the first 12. The plates, on this occasion, were
tied up in bundles of a dozen, and the accession

of magnetic power was observed on every addition of two such bundles.

Series of 72 hard Plates.						Series of 192.		
No. of Plts.	Mean Power.		No. of Plts.	Mean Power.		No. of Plts.	Mean Power.	
	Devia- tion.	Tangent.		Devia- tion.	Tangent.		Devia- tion.	Tangent.
1	3·18	58	14	32·25	635	12	26·42	503
2	6·23	112	16	35·5	702	24	41·8	873
3	9·41	171	18	37·15	760	48	54·14	1388
4	12·10	216	20	39·15	817	72	60·16	1751
5	14·38	261	22	41·15	877	96	63·20	1991
6	17·15	311	24	42·38	921	120	66·0	2246
7	19·22	352	30	47·2	1074	144	68.32	2543
8	21·40	397	36	50·52	1229	158	70·21	2801
9	23·50	442	42	53·17	1341	192	72·6	3096
10	25·37	479	48	55·30	1455			
11	27·50	528	60	58·32	1634			
12	29·17	561*	72	61·19	1828			

* The first 12 plates of the series of 72 were combined under a
pressure only of 2 pounds, the remainder under a 3-pound pressure.
The plates, however, being uneven, were far from being in general
contact.   When the 48 plates, producing a deviation of 55° 30', were
compressed, by being tied tightly with twine, the deviation fell to
53° 15'; but they recovered their power, when restored to their former
state of pressure, to the extent indicated by the deviation, 54° 38'.

It has been shewn, in the Results of the investigations on the powers of combinations of
magnetised plates of *tempered* steel, in contact,
(Part II. chap. I.), that such combinations, if
numerous, soon cease to be very effective; so
that no extent of additions would more than
double the power of the first six plates of the kind
there referred to.   But here, in the use of hard
plates, we have a remarkable and satisfactory

M

improvement; for we find the power of the first six plates actually doubled by 14, trebled by 24, quadrupled by 36; whilst the power of 192 plates is nearly ten times as great as that of six!

Comparing, again, the powers of these hard plates, in combination, with those of bars, tempered as they usually are in the manufacture of magnets in this country, we find equally remarkable and satisfactory results.

Not to take the case of magnets of different lengths, where the comparison becomes less conclusive, I shall here only adduce the proportions of the powers of a set of bars, constructed expressly for this comparison, of the same length and breadth as these dimensions of the hard steel plates.

These bars were made of " blister steel," and consisted of one pair of the thickness of 0·28 inch, weighing 1·762 pounds each; and of another pair of the thickness of 0·47 inch, weighing, on an average, 3·006 pounds. These thinner bars were each equal in mass to a little more than eleven of the hard plates; and the thicker to about twenty.

In the first comparison, the bars were only tempered at a blue, in a very short extent at the ends, (the whole of the intermediate portions being soft,) and in that state the power of the *thin* bars, separately, was to that of an equal

mass of hard plates, as 1 to 2 64. But the thick
bars, being of inferior quality, not having undergone the same hammering (see p. 84), had only
a magnetic capability, compared with that of an
equal mass of hard plates, in the proportion of
187 to 817, or as 1 to 4·4 nearly. The power
of the two thinner bars together, compared with
a correspónding weight of hard plates, was as
1 to 3·9: and of the two thick bars as 1 to 6·0.

The bars were then made *hard* at the ends
(without being reduced or tempered) for an extent of about three inches, when their magnetic
capabilities were found to have improved; so
that each thin bar, compared with a like weight
of hard plates, now exhibited the ratio of 1 to 2;
and each thick bar of 1 to 3·4. Tried in combination, the ratio of power of the pair of thin
bars, with an equal mass of hard plates, was as
1 to 2·7; and of the pair of thick bars, as 1 to 4·5.

In the next comparison, the hardness of the
ends of the bars was reduced to a blue temper,
when the comparison, with equal masses of hard
plates, gave the following ratios:—Each thin
bar, as 1 to 2·3; the pair of thin bars, as 1 to
3·2. Each thick bar, as 1 to 3·6; and the pair
of thick bars, as 1 to 4·7.

For a final comparison, the bars were ordered
to be made as hard as possible; but this order
was only accomplished with the thinner pair—

all the efforts of a clever artisan having failed to
harden the others. The result with the thinner
bars, however, was satisfactory—each bar exhibiting a much nearer advance towards the
powers of the hard plates ; the proportion of
equal masses now being as 1 to 1·46, with one bar,
and as 1 to 1·62, with the two bars combined.

We thus find the superiority of the magnetic
powers of combinations of hard plates, over those
of bars tempered or hardened in different degrees, to be most decisive. Whilst the power of
these *hard* plates, individually, was (in accordance with former experiments and with theory)
found to be *inferior* to that of *tempered* plates ;
that inferiority, it is obvious, only belongs to a
very limited mass. For inasmuch as the deterioration of the hard plates, by the violence of
combination, is so much less considerable than
it is in softer or tempered plates or bars,—the
proportional power in the hard plates, as so combined, rapidly gains upon that of combinations
of *tempered* steel, so as, after a series, perhaps, of
six plates, to come to an equality of power with
the best capacities of steel of any other temper.
Beyond that proportion of combination, the comparison becomes wholly in favour of the hard
plates, and goes on, in an improving ratio, to
the utmost extent to which addition to the combination can be usefully made.

The diagram No. 3, Plate III , will satisfactorily represent to the eye, the comparative
powers of combinations of hard steel plates, and
those of single bars, or pairs of bars, as tempered
at the ends only, in the ordinary manner of magnets, and also as hardened throughout. The
dotted portions of the curves are only inferential;
but they are abundantly borne out, as proximate
powers, by analogy with the curves yielded by
the powers of large combinations of bars of other
dimensions.

The series of successive powers, it may be
observed, exhibited by combinations of hard
plates (as given in the preceding table), is not
quite so regular as that heretofore obtained
from combinations of *tempered* plates, and this
for one very obvious reason: the imperfect and
irregular nature of the contact obtained in these
unavoidably uneven plates. Besides this inequality in the measure of contact in the smaller
bundles of plates,—a similar condition was inseparable from the various combinations of the
different bundles; so that when, at certain limits,
the whole were strapped together, the increase of
pressure necessarily occasioned a diminution in
the power of the whole fasciculus, and a conse-
quent inequality in the series. After the last
result had been obtained, the whole mass, of 192
plates, was formed into a square prism of equal

sides : the magnetism, then, being more con-
densed, the power was considerably reduced.
So also, some of the other combinations, when
unduly pressed together, were reduced beneath
the general average.

From this diagram—in which the abscisses
are on a scale of just five-ninths of those of
diagram No. 2, and the ordinates of one-third
—we may readily perceive the amazing advantage, in large combinations, of the magnetic
capabilities of hard plates, over plates or bars
otherwise tempered.   The ultimate gain of the
hard plates, over a combination of bars of
0·28 inch thick, tempered along a considerable
extent at the ends, appears, by the proximate
curve, to be in the ratio of nearly seven to one;
or compared with a series of similar bars, tempered in the manner in which compound magnets
are usually prepared in this country, the ratio
would be still higher.   And comparing these
powers of the hard plates, with the power
of the same bars when hardened in a like
manner, as far as the steel was susceptible of
hardness, we find (by reference to the middle
curve) that the fasciculus of plates still has the
advantage in the proportion of, probably, nearly
two to one.   Had, however, the combination of
hard plates been carried farther, it is quite
obvious, from the mere inspection of the diagram,

that the differences, in their favour, would have
been increasingly conspicuous ; for whilst the
bars of the ordinary kind of temper had long
ceased to gain anything worth notice, by additions of like bars to the mass,—as shewn by
the horizontality of the extension of the lower
curve,—the curve, expressive of the power of
the hard plates, is still found to rise at a considerable angle with respect to the direction of
the line of abscisses.

It has been remarked, that the proximate part
of the curves (distinguished by dotted lines) has
the support of analogy derived from actual experiments with some large series of bars and plates
of other dimensions: so that the direction and
elevation of the lower curve especially, cannot,
I think, be far from the truth. The *probable*
accuracy, however, it is to be observed, only
relates to results with corresponding, or medium,
qualities of steel; for it should be borne in
mind, that the statement often made—" that the
magnetic qualities of different denominations of
steel, if of ordinary goodness, are not essentially
various"—is most fallacious, as *very great differences* will be found, as before shewn, in the
magnetic powers of the several descriptions and
qualities of steel. Hence, none of the curves
of these diagrams, if applied *generally* are to
be considered otherwise than proximate, as to

their elevation, or as to the measures of their ordinates.

As to the diagram No. 3, the curve, representing the powers of hard plates, has been derived from experiment with steel of ordinary quality, so that, were the best cast-steel substituted for a similar experiment, a very large increase would doubtless be given to the elevation of the curve.

(2). *As to hard* bars *in combination.* — The great gain in magnetic power obtained by the use of hard *plates*, rendered it quite certain that a gain, somewhat similar, would result from the thorough hardening of the *bars* designed for magnetic combinations. The measure of gain, however, was matter of experiment.

A set of bars well adapted for such experiment was that repeatedly referred to in chapter i., of this part of these Investigations, consisting of a series of similar bars, ten in number, 13·8 inches in length; 1·1 inch in breadth, and 0·255 in thickness. The average weight of these bars was 7462 grains, and the total weight of the ten, 10·66 pounds.

They were orginally made for a pair of compound straight-bar magnets, five bars in each, and tempered in the usual manner at the ends only. Their powers were always moderate; and the apparatus, which they formed, was found

soon to lose much of its energy Having carefully magnetised the whole set, determined their
separate powers, tested their respective qualities,
and tried their powers in combination, on every
addition of bar by bar to the mass, up to the
completion of the series,—I then sent them
to a manufacturer in steel, at Sheffield, to be
thoroughly hardened.

On their return, though the hardness of the
bars was neither complete, nor equable; yet a
most decided and satisfactory improvement was
found to have taken place in their magnetical
capabilities. The comparison of their respective
powers in the two conditions, as to hardness, will
be the best exhibited in a tabular form:

TABLE

TABLE of the Magnetical Powers (Maximum Capacity and Tenaciousness) of a Set of Ten Bars of 13·8 Inches in length, in Two States of Hardness.

[The distance from the Compass was Two Lengths, or 27·6 Inches.]

Nos. on the Bars.	Weight in Grains.	I.—Bars tempered only at the Ends.				II. The same hardened throughout at a low red heat.			
		Highest Deviation separately.	Deviation when *reduced* by combining 5 Bars.	Powers in Combination.		Highest Powers separately.	*Reduced* by Combination of 10 Bars.	Powers in Combination.	
				Mean.	Tangent.			Mean.	Tangent.
I.	II.	III.	IV.	V.	VI.	VII.	VIII.	IX.	X.
III.	7575	17·10	12·12	17·10	309	18·5	12·20	18·5	327
VIII.	7480	16·0	11·0	20·0	364	19·8	12·4	30·5	579
X.	7380	15·5	10·5	21·10	387	19·5	14·47	36·20	735
VII.	7500	15·20	9·35	22·55	423	19·56	14·40	41·50	895
V.	7270	16·40	8·30	25·30	480	19·12	13·34	44·40	988
I.	7500	15·55	8·5	28·30	543	18·20	13·44	47·7	1077
IV.	7550	15·50	8·0	30·10	581	18·0	5·45	[omitted.]	[omitted.]
II.	7300	15·30	6·55	31·30	618	18·20	14·23	49·55	1188
IX.	7510	12·20	1·10	30·0	577	17·30	8·33	51·20	1250
VI.	7610	11·0	0·50	28·45	549	14·22	—2·40	[49·10]	[1157]

The effect of hardening this set of bars, was thus shewn to be in exact correspondence with what had been anticipated. Comparing columns III. and VII., we perceive the general gain of magnetic capacity, by hardening, in the bars individually; and comparing columns VI. and X., we see the still further gain obtained under the respective combinations, bar by bar, up to the amount of the whole series.

A similar experiment was made with the bars of a large compound magnet of the horse-shoe form, and with an analogous result. These bars were nine in number, and had been moderately hardened, in the usual way, only at the ends or poles. They were composed of the ordinary quality of steel, and weighed, on an average, 2·56 pounds, or, in the whole, 23·05 pounds. In the original state of the instrument they had a lifting power, separately, when most highly magnetised, of about 14·3 pounds; but this power, after the conductor had been once removed, fell to that of 8·9 pounds.

As it would have been a very difficult task to have hardened these bars throughout, without change of form—I got them hardened, at a bright red heat, in all parts except at the bend, which, being left soft, enabled the artisan to adjust them, pretty nearly, to their original shape.

With this measure of hardening, an improvement, fully as great as what I had anticipated,
was found to have taken place in their magnetical properties. Their separate lifting powers,
indeed, *before* the removal of the conductors
which had been attached to them, severally,
during the magnetising process, was not greatly
augmented—the amount of their powers, as tried
separately, having only increased from 124 to
134 pounds. But *after* the removal of the
conductors, the comparison of their subsequent
powers indicated an improvement in the proportion of the advance from 80 pounds to 120
pounds,—the former being the amount of their
separate lifting powers after the removal of
the conductors in their *original* state, as to
tempering; and the latter in their *hardened*
condition.

When the power of the whole series of bars
was tried in combination—an equally satisfactory improvement was determined. In the
original state of the complete magnet,—after
being highly magnetised, and carefully put up
whilst each bar was defended by its own conductor, and these separate conductors gradually
displaced by sliding on the large conductor of
the entire instrument,—the power sustained by
it was gradually increased to 70 pounds before
the conductor was pulled off: but, on a second

trial, the lifting power was found to have fallen
to about 30 pounds. Now, however, after the
change in its degree of hardness, it was found
to sustain (under the great disadvantage of an
uneven surface where the conductor was applied)
considerably above 80 pounds; and after the
conductor had been separated, several different
times, it still readily sustained 80 pounds, or
upward!

Hence, the very manifest improvement of the
instrument. First, a considerable increase in
its sustaining power before the removal of the
conductor; and, secondly, an improved degree of
permanency of power, such as that indicated by
a comparison of the powers in the two different
states, when once the conductor had been
removed. In the original state, the sustaining
power fell, on the removal of the conductor, from
70 to about 30 pounds; in the improved state,
the power only fell from, perhaps, about 85 to
80 pounds, or scarcely so much!

Finally, under this head of Investigations, I
may mention the effect of hardening on the
series of 8·7 inch plates—which had been
subjected to such a variety of treatment in tempering—for the purpose of following out the
inquiries in section I. of this series. I took eight
of those which had been thoroughly softened in

the fire (p. 144), and annealed them at a bright red heat, in cold water. They were found to be extremely hard, and as brittle as glass. Their magnetic capacity, separately, was now reduced from a deviating power (at two lengths from the compass) of near 10°, in their original state, at a spring temper, to about 8°. But their power in combination had proportionally increased—the eight plates now exhibiting a power of 30° 30′, whereas twenty-four plates, in the original state, as to temper, only had a power of 28° 41′.

In the diagram No. 4, the powers of these eight hard plates are represented, and carried forward by the continuance of the curve, to the probable result of the whole of the original series of twenty-four. The lowest curve in the diagram represents (as projected in like manner) the powers of the same plates when thoroughly softened.

RESULTS.

Of the General Results of the foregoing Investi
gations on the powers of magnetised steel plates
or bars, *as affected by differences in the degree
of hardness*, or nature of temper,—I proceed to
select a summary. But in order to the extension
of results, having a most important bearing,
practically, on one of the leading objects of this
work—the improvement of magnetical apparatus—I shall here avail myself of certain
important facts (being in strict conformity, in
analogy, as to the effects of temper) elsewhere
derived, and especially of those elicited by the
Investigations in Part I., as exhibited to the eye
in the diagram in Plate II.

1. That the relative powers of *combinations* of
magnetised plates or bars of steel, as well as
those of single pieces,—are greatly affected by
differences in the state of the steel, both as to its
quality and as to its temper.

By the *powers* of magnets is to be understood, not the
capacity for magnetism, as indicated by the force
required, in the first instance, to remove the conductor
by which, during the magnetising process, the magnetic

arrangement had been sustained, but the magnetic energy which remains in a bar when detached from the conductor, or other sustaining magnets.

The effects of the *quality* and *denomination* of steel, on its magnetic properties, have, to a certain extent, been shewn in Part I. chap. VII., and the particular cases noticed at pages 46 and 84.

The general effect of *quality* becomes more and more important, as the mass, or combination employed, is enlarged. Hence, *inferior* qualities of steel, however energetic small magnets, constructed out of them, may appear, are incapable of very extensive combinations with advantage, or of being usefully employed, as magnets, in large and heavy bars.

Hardness of the steel operates in a similar manner as *quality* on the magnetic properties of large masses; so that increase of hardness greatly improves the capacity for *sustaining* the magnetical arrangement or condition, where, on account of the size, the tension is great.

2. That various degrees of hardness have an influence on the magnetic capacity and energy of steel, differing, both in the *nature* and *quantity*, or proportion, with the magnitude of the masses employed; so that that kind of tempering, or degree of hardness, which may exhibit superiority with a certain mass, may be found greatly inferior in other magnitudes.

This result will be very apparent, if we merely inspect the relation of the curves in either of the diagrams, No. 1, or No. 3. The first diagram shews, very strikingly, the change which takes place in the relative powers, as to directive energy, in magnets of different degrees of hardness. Thus comparing the effects of the four distinctive kinds of temper there indicated, on the magnetic capabilities of very thin plates; such as, for instance, on single plates of 6 inches long, and of the weight only of 100 grains,—we find, that the order of energy, commencing with that of the highest, is in favour of— 1st. Tempering at the ends; 2d, tempering throughout; 3d, being left soft; and 4th, being made quite hard; or, to use the distinguishing letters there employed, in the order E. T. S. H.—Comparing, next, the powers of plates of 200 grains weight, the order is found to be changed to T. E. H. S.—Comparing again the powers of plates of 300 grains, we have the order T. H. E. S. —And comparing, finally, the powers of the same sets of bars in masses of 400 grains, we have the order, never afterwards altered, except as to quantity, or proportions, H. T. E. S.

Analogous results, as to the respective powers of *cast steel* bars or plates of different degrees of hardness, are derived from the inspection of diagrams Nos. 3 and 4, and also from the experiments with busks of various degrees of hardness described in section II. of this chapter.—But more particular reference to these cases is reserved for the illustration of the next proposition in these Results.

With regard to the *proportions* of the magnetic powers of bars of cast steel, or combinations of cast steel plates, of different degrees of hardness, we find a change with every alteration in the masses compared. Thus, with reference to the comparative powers of magnetised steel bars, as exhibited in the first diagram, we find the ratio, unequated, of the series of bars No. I. (of about 3000 grains weight) to be 100, 76·6, 58·3, and 40·9, in the order of the several degrees of hardness H. T. E. S.;

whilst the ratio of the corresponding series of bars
No. ii. (being about 1400 grains weight) appears to
be—100, 79, 67, 49; and the ratio of series No. iii.
(of about 700 grains)—100, 87, 75, 56, nearly.

These proportions, as well as the places in the line of
abscissæ at which the different curves cross each other,
are only proximate. For in case of the degree of hardness of any of the sets being different, some alterations
would be made both as to the proportions of the powers
of the different descriptions of bars, and as to the
places of the intersection of the curves.

3. That the pre-eminence of *partially tempered*
or *slightly hardened* straight-bar magnets, for the
magnetic condition, belongs only to very limited
masses, — whilst in fixidity, or permanency of
power, the softer magnets are always inferior.

The first part only of this proposition needs to be illustrated here—the second belonging more particularly to
the investigations of a subsequent chapter.

It has just been shewn that, however in certain small
masses a *moderate degree of hardness* may have an
advantage in the first instance, as to magnetic energy,—
on the comparison of bars of six inches in length, and
half an inch in breadth, it was found, that this occa-
sional pre-eminence in any inferior measure of hardness
gives place to the condition of the greatest hardness,
in the particular case investigated, whenever the weight
of the bar (being of best cast steel) exceeds 400 grains,
or the thickness one-fifteenth of an inch.

In the case of the experiments described in section ii.
with the variously tempered busks of fifteen inches in
length, analogous results were obtained.

Comparing these plates, as to their relative magnetic powers, *singly*, we find (table at page 155) that the most energetic were those of a spring temper throughout—Nos. vi. and vii.; next in energy were those in the soft, or " raw state," No. i.; and the feeblest were the very hard set, No. ii. But when three of each kind were combined together, the hard busks, which, individually, had only two-thirds as much energy as those that were of an elastic temper, had now *more* than three-fourths of the power of these; whilst they had already advanced beyond, or nearly to equality with, several of the other kinds.

Carrying the comparison still further, which other experiments not herein detailed enable me to do, it was found that, after a combination of about eight plates only, or a total thickness of cast steel of about the fifth of an inch, the thoroughly hard busks had an universal pre-eminence.

In the comparison of *bars* in combination, in sets of different degrees of hardness—a similar limit to the superiority of those moderately hardened, as to the first energy received, was experimentally determined. In the case of the two-feet bars described at page 101, the powers of which were determined in three states of hardness—there was found a progressive increase in correspondence with the increase of hardness; and in the case of the sets of 13·8 inch bars and of those of the large horse-shoe magnet, referred to in the last section of this chapter,—each bar, separately, gained somewhat in energy by being hardened, and when tried in combination exhibited a large accession to their original power, as tempered in the ordinary manner.

In diagram No. 2, the powers of perfectly hard plates of cast steel, are compared with those tempered like cutlery throughout. At the 12th plate the inferiority of the hard plates disappears; and beyond that number they assume their usual pre-eminence over those of less hardness when in similar contact.

4. That, in combinations to a small extent of magnetic power, a moderate degree of hardening throughout will yield effective or superior energy; and that measure of energy will, under certain conditions, be increased by slightly heating or reducing the middle of the thoroughly tempered plates.

This proposition, as to the first part of it, has already had abundant verification. The second part of it, which is analogous to a conclusion arrived at by Captain Kater, is derived from the experiments detailed in the first section of this chapter, where the measure of the apparent reduction, and the proportion of gain in magnetical capacity, have been already considered.

The advantage yielded, however, by this slight reduction in hardness, to the immediate magnetic energy of the particular set of plates referred to at pages 134 to 138, was found, as the extent of combination advanced, to become proportionally less and less considerable. The case, indeed, as determined by Captain Kater, was limited to that of thin plates, employed separately as compass needles; and here the same result is found to be correct in combinations of plates to a very considerable extent in number, the energy being but moderate.

This circumstance, as has been shewn in the text of the foregoing chapter, occasioned me, at first, much embarrassment, as being, apparently, at variance with the law so extensively verified respecting the general advantage of hardness for magnetic power in large combinations of steel plates or bars. But, after a variety of investigations for the solution of the apparent discrepancy, it was found to be resolvable into the general law, the difficulty having arisen from want of regard to the actual weakness of the combination. Though the plates were so numerous (being twenty-four in number), yet they were so little energetic in combination that the whole series did not double the power of the first two,—the total energy, indeed, not exceeding what might be obtained in a single hard plate of shear steel, or tempered plate of cast steel of, perhaps, 1500 to 1800 grains weight.

Had, therefore, the combination been carried to a high degree of energy, it is obvious that the powers of the harder series would have become superior. For the harder series was found so to gain upon those that were reduced, as the mass increased, that it would have come, no doubt, to an equality, if the number of plates had been augmented to thirty-two, and beyond that number would have obtained the pre-eminence. See diagram, No. 4.

5. That whilst the order of superiority in magnetic plates or bars of *like* form and length, but of *different* degrees or measures of hardness, varies, within certain limits, as the mass is diminished or enlarged; yet, the limits of the degrees of hardness which are the most effective for *all practical uses*, may be considered as comprised betwixt a brittle hardness, like that of files, and that of an elastic or spring temper.

The relation to each other of the various curves in diagram No. i. (plate 11), as they appear near their commencement, may be appealed to as sufficient to support the truth of this statement as a general practical proposition. For herein it plainly appears, that, as to all essential or material advantages, in any particular kind of tempering or degree of hardness—the choice mainly rests between the degrees of hardness indicated by the curves T and H, or the intermediate modifications.

As to the case of plates or bars perfectly soft (S), we find that this condition is unequivocally inferior; and as to the case of plates or bars tempered only at the ends (E), that the superiority in very thin plates which, within small limits as to mass, it appears to possess, is only to a trifling extent;—so trifling indeed, as, when taken in connexion with its disadvantage from the want of fixidity, wholly to set aside this mode of tempering, so generally adopted in the construction of compass needles and other magnetical instruments, as unscientific and bad.

6. That the sustaining property of hardness, on the magnetic arrangement, is such, as to render a considerable extent of hardness generally advantageous, for magnetical apparatus; whilst in very large combinations of straight-bar magnets, or very heavy bars, the greater the hardness the more powerful will be the magnet.

For particular purposes, where a light needle, or plate, may be required—a considerable reduction of the ultimate hardness of which *cast steel* is capable, will be advantageous to its primary energy, and general utility, though the fixidity be not so great. And in magnets of the horse-shoe form, a reduction of *extreme* hardness is generally beneficial. But in all larger masses, in straight-bar magnets, a temper approaching to extreme hardness, becomes in every way the most advantageous condition.

Hence, whatever may be the advantage to the individual magnetic capacity of thin steel plates, by a reduction of hardness; the superior strength or retentiveness of hard plates, eventually causes them greatly to exceed in power, under energetic combinations, those of any other kind of temper.

In the case of the fifteen-inch plates, or busks, very numerously combined, we find that, compared *separately*, the very hard plates had scarcely two-thirds of the power of those of an elastic temper; but when combined to the extent of 192, the hard plates exhibited a degree of energy, probably six or eight times as great as that of which similar plates would be susceptible of a spring temper.—See diagram No. 3.

7. That the most favourable conditions for the construction of magnetic apparatus, by the combination of several bars or plates, require *similarity* of quality and denomination in the steel, a reasonable correspondence in the magnitude or mass of the individual plates or bars, and similarity of temper or hardness.

The first and last of these requisites for making the best use of any series of bars or plates in combination, are proved to be essential by simple inference from the general results of investigations already given, whilst the processes described for testing the quality and hardness enable us very easily to make the proper selections.

The practical methods, in detail, however, will remain to be described in a subsequent part of this work.

The desirableness of the second condition, that of correspondency of mass in the plates or bars combined, may not appear to have been yet determined; though when we consider the impracticability of giving equal degrees of hardness to thick and thin bars, and of equally developing the magnetic energy in thick masses, as we can in thin plates, the requirement may be seen to be one of considerable importance.

The above Results, it should be observed, relate, essentially, to one particular *denomination* of steel—that of *cast steel*—the relative properties of this and various other denominations of steel (both in single bars, and in various combinations, as well as differences of temper), forming the subjects of investigation in the next succeeding chapters. It should be further borne in mind, that the comparative powers, herein treated of, likewise relate, not to all the possible changes producible by every variety of hardness of which steel is susceptible, but, only to such general divisions or species of hardness as have heretofore been designated by the characteristic letters S. E. T. H.

CHAPTER IV.

ON THE RELATIVE POWERS IN COMBINATION, AND SEPA-
RATELY, OF HARD PLATES OR BARS OF STEEL OF
DIFFERENT DENOMINATIONS AND QUALITIES.

———

THAT a striking relation exists betwixt the *quality*
of steel, and its magnetical properties, has been
shewn, I trust I may say, demonstrated, in
chapters VII. and VIII. of the first part of this
work. On the other hand it has been shewn,
that certain *differences*, probably characteristic,
occur in the magnetical powers of steel of different *denominations*—such as "cast," "shear,"
"double shear," "blister steel," etc., as the
several kinds are designated with reference to
the manner in which the metal is treated after
being "converted"—though the several varieties
of steel, as to their denomination in commerce,
may have been converted out of the same *quality*
of iron.

The experiments, however, heretofore recorded on both these important points, were of too limited a nature to be quite conclusive for the solution of the important problem,—as to the best quality or kind of steel for *all* magnetical purposes, or, rather, for each magnetical purpose specially. For whilst the fact was sufficiently clear, that the *best qualities* of steel have the highest magnetical powers, and, that, for certain magnetical purposes, and powerful magnetical combinations, the best *cast* steel exhibits a decided superiority,—it by no means followed, that *cast* steel as to denomination (not quality), would maintain that relative superiority, in all the different forms, and masses, and degrees of hardness in which the bars or plates of magnetical apparatus might require to be constructed. Nor did it necessarily follow that the results obtained with thick bars, (Part I. chap. vii.)— because of the difficulty of hardening and magnetising them throughout their mass—would, in all cases, be found, especially in proportion of magnetical energy, correspondent with results obtained by combinations of thin plates.

Hence, I thought it needful to institute the additional series of experiments of the class indicated in the title of this and following chapters. And the importance and satisfactoriness of the results amply fulfilled the anticipations I had entertained.

In the series of experiments embraced in this chapter (with the view of special application to the case of compass needles of larger dimensions than 6 inches) I adopted generally the length of 7·5 inches, for the steel plates, with the breadth of 0·75 inch, and a thickness, for the most part, of about 0·08 inch.

The thickness, however, in the several series' designed to be similar, was found, from circumstances which I could not easily control, to vary sometimes more than I could have desired.

One general method of hardening was adopted in the whole of the plates referred to in this chapter; and all the plates, but one set, passed, in their corresponding stages of rolling, forging, grinding, and hardening, through the hands, I believe, of the same artizans. The mode of hardening adopted, was by plunging the plates, when at a bright red heat, into a strong brine. The hardness thus yielded was equivalent to that of files, and was not allowed in any way to be reduced, until the whole of the experiments described in this chapter had been completed.

For the facility of reference and discrimination of the several denominations, qualities, and degrees of hardness of the steel, etc. employed in the investigations comprised in this and the succeeding chapters, the following series of characteristic letters have been adopted.

Characteristic letters of *Denominations* of steel and iron; the *best qualities*, referred to indefinitely, being indicated by the letters in italic capitals.

C. Cast Steel.

S. Double Shear Steel.

B. Blister Steel.

CL. Cast Steel known to be from hoop-L iron.

CS. Stubs' Cast Steel.

SS. Stubs' Double Shear Steel.

I. Iron.

CI. Cast Iron.

M. Steel used at the Mint.

Characteristic letters of the *Qualities* of steel in small (or lower-case) italics.

f. Steel from a Foreign Iron, not being of the very best, but of a good quality.

fc. Steel from a common kind of Foreign Iron.

e. Steel from undescribed kind of English Iron.

b. Steel from best Bradford Iron.

Characteristic letters, in Roman capitals, as partly employed in the diagram, Plate II., etc., indicative of the temper, or degrees of hardness.

H. Quite hard; hardened the same way, and to the same extent, as files usually are.

T. Tempered by being reduced to a blue, or spring temper, equably throughout.

E. Tempered at a blue, but only at the ends of the bars, the intermediate portions being soft.

S. Soft; just as received from the rolling-mill or forge.

HT. Intermediate, betwixt hard and tempered.

TS. Intermediate, betwixt tempered and soft.

AB. Annealed in boiling linseed oil.

A. Annealed in hot oil at some specified temperature.

SECT. I.—*Of the Magnetical Powers, in combination, of Steel Plates (7·5 inches in length), of* DIFFERENT DENOMINATIONS, *but of* THE SAME ORIGINAL QUALITY *of iron.*

Steel of the several denominations of " cast," " shear," and " blister," constituted the chief varieties in the investigations of this section,—the specimens for which were kindly furnished me, for the most part, by Messrs. Stubs, of Warrington, whose tools and files have obtained such just celebrity.

1. Experiments with steel of *different denominations,* converted out of a *peculiar steel (CS)* used by the Messrs. Stubs in the manufacture of files, etc. The basis of this steel is all foreign iron, smelted with wood-charcoal, of the first qualities.

In each case the plates were similarly magnetized by the process K S, upon the large compound magnets of 192 hard plates. The two-feet hard bars were generally found adequate for the development of the highest magnetic condition of which the plates were susceptible; but to secure the maximum energy, without chance of failure, the more powerful magnets were preferred. The deviations produced by the plates, or combinations of plates, were all determined on the trial-board, and at the distance of 15 inches, or two lengths from the compass.

No. of the Plates.	Maximum Power Separately.	Power in Combination.		Power after Combination.
		Deviation.	Tangent.	
I.	II.	III.	IV.	V.

I.—*Cast Steel*: Average Weight, 986 grains.
[Tested under a Combination of 36 Plates, or Deviating Power of 71° 47′.]

C. H. 1	23·0	23·0	424	14·6
2	18·52	33·32	663	11·0
3	18·42	41·9	874	11·24
4	21·24	47·30	1091	14·0
5	19·20	51·15	1246	9·0
6	20·40	53·44	1364	13·6
Mean	20·20	. . .	371	12·6

II.—*Double Shear*: Average Weight, 890 grains.
[Tested under a Deviating Power of 71° 47′.]

S. H. 1	23·40	23·40	438	3·45
2	23·37	35·20	709	4·27
3	23·30	41·27	883	4·45
4	21·15	44·25	980	6·27
5	21·45	47·20	1085	5·26
6*	22·45	49·0	1150	5·0
Mean	22·45	. . .	419	4·58

III.—*Double Shear*: Average Weight, 630 grains.
[Tested under a Combination of 28 Plates, or Deviating Power of 54° 40′.]

S. H. 1	17·0	17·0	306	4·15
2	16·40	25·0	466	2·28
3	16·30	29·28	565	11·6
4	16·45	31·45	619	5·42
5	16·0	33·46	669	9·32
6	16·53	35·15	707	·6·32
Mean	16·38	. . .	299	6·36

* This is an assumed Plate to make up the Series, five Plates of this kind only being obtained.

No. of the Plates.	Maximum Power Separately.	Power in Combination.		Power after Combination.
		Deviation.	Tangent.	
I.	II.	III.	IV.	V.

IV.—*Rolled Cast Steel:* Weight, 590 grains.

[Tested under Deviating Power of 54° 40′.]

C. H. 1	11·12	11·12	198	7·2
2	12·0	19·54	362	6·54
3	11·45	26·52	507	6·0
4	12·0	32·40	641	7·16
5	12·30	36·0	727	8·18
6	12·35	39·0	810	8·42
7	15·0	41·10	874	8·8
8	14·40	44·0	966	9·3
9	15·46	46·0	1036	8·44
10	15·22	47·0	1072	8·40
Mean	13·17	. . .	236	7·53

The "double shear steel" of series III. was supposed, by the persons by whom supplied, not to be highly carbonized. It had been prepared specially for making *springs*, rolled into sheets, and well hammered, when cold, before being hardened.

The "rolled cast steel," series IV., had been prepared for *saw-blades*, and was of a description denominated "very mild."

For the correct comparison of the powers of these four descriptions of steel, an allowance, it will be observed, requires to be made for the differences in weight of the respective series. This may be effected, proximately, by the use of the diagram, Plate II., wherein the powers of plates of 7·5 inches will be represented by those plates of 6 inches, by taking the former in the ratio of the cubes of their lengths, viz. as $7·5^3$: $6·0^3$, or as 421·875 to 216, or sufficiently near at one-half of their weights.

A correction for the difference of *hardness* is likewise requisite: for it may be perceived, by comparing the loss of power produced in series No. I. by the combination of 36 plates, having a deviating power altogether of 71° 47', with the loss sustained by series No. II., under the same severity of testing, that the double-shear steel was reduced greatly below the cast steel, indicating a much less degree of hardness. The loss of power in the *shear* steel plates, by the severity of the test to which they were subjected, was, on average, nearly 80 per cent ; whilst the loss sustained by the *cast* steel, under the same severity of trial, was only about 42 per cent. Hence the correction *for weight*, to be added to the mean powers of the plates of set No. II., separately, for a proper comparison with those of No. I., would, in due respect of hardness, be according to the ratio of a curve below that of H, in the diagram, and probably above that marked T. The augmentation betwixt $\frac{890}{2}$, the weight of plates of series No. II. and $\frac{986}{2}$, that of No. I., would in this case amount to 10 on the

line of tangents, which being added to 419, the
tangent of 22° 45', or the mean maximum power
of the plates of series No. ii. separately, would
give 429, equivalent to a deviation of 23° 13'.
If to the two series, Nos. iii. and iv., the
correction for deficiency of weight were applied,
with a due regard to their respective curves for
hardness, about 80 would have to be added to
the tangent of 16° 38' of No. iii., and about 100
to that of tangent 13° 17' of No. iv.,—giving to
No. iii., on a mean of the maximum powers of
the several plates of the series raised to a correspondence with the weight of No. i., a deviation
of 20° 46', and to No. iv., of about 18° 35'.

The numbers, however, thus obtained are, at
the best, but proximate; and where such a variety
of circumstances affect the general results, these,
within certain limits, must be uncertain. At
the same time, some very important practical
facts are abundantly developed; but as these
will be rendered more conclusive and satisfactory
by the experiments yet to be described, they
may be with advantage reserved for the summary of propositions after our usual manner,
under the title of " Results."

A deduction or two, however, of some importance in the manufactures out of steel, as well
as in practical magnetics, may be now noticed.
Comparing the power and retentiveness of the

plates used in series No. i., with those of No. ii.,
we find, by the much less retentiveness of the
latter, (col. v.) that the hardening of *different
denominations* of steel in the same way, and at the
same heat, produces very different effects, as to
the degree of hardness. Thus the process which
yields to best cast steel the highest degree of
hardness, H; yields to double shear steel of
similar quality, a hardness but little superior,
apparently, to that marked T. The difference, it
would appear, mainly arises from something in
the structure of the metal,—the mode of manufacture producing a peculiarity in the arrangement of the ferruginous particles, so as to modify
their respective susceptibilities for hardness.

Another deduction, from a comparison of the
corrected powers of the series iii., of shear steel,
with those of the similar denomination, No. ii.;
as also from a comparison of the cast-steel, No. iv.
with No. i., is of some importance for the further
verification of the principle of testing. Thus, we
find, the mean powers of the plates of each of
the four series, when corrected for weight and
hardness, to be—No. i. (cast-steel) 20° 20', and
No. iv. (cast-steel) 18° 35 ; No. ii. (shear-steel)
23° 13, and No. iii. (shear-steel) 20° 46'--shew-
ing an inferiority, in magnetical qualities, of
the series iii. and iv. Now such inferiority
was fully anticipated from the knowledge of the

fact of their being in an inferior degree carbonized—a condition designedly given, because of
the toughness required in the peculiar purposes
for which these two steels were designed—viz.
for *springs* and *saws.*

2. Experiments with steel of *different denominations*, converted out of *best Swedish iron*,
marked hoop-L.

No. of the Plates.	Maximum Power Separately.	Power in Combination.		Power after Combination.
		Deviation.	Tangent.	
I.	II.	III.	IV.	V.

V.—*Cast Steel:* Average Weight, 952 grains.
[Tested under a Deviating Power of 71° 47'.]

CL. H. 1	20·33	20·33	375	13·15
2	19·18	33·48	669	11·45
3	19·30	42·18	910	12·0
4	20·40	48·45	1140	13·10
5	20·37	52·44	1314	13·12
6	22·2	55·30	1455	13·13
Mean	20·27	. . .	373	12·46

VI.—*Cast Steel:* (2nd Series of the above) 952 grains.

CL. H. 7	19·0	19·0	344	11·28
8	19·36	32·35	639	11·30
9	19·45	41·33	886	12·57
10	19·30	47·22	1086	12·22
11	20·15	51·35	1261	12·37
12	19·30	54·20	1393	12·22
Mean	19·36	. . .	356	12·13

| No. of the Plates. | Maximum Power Separately. | Power in Combination. | | Power after Combination. |
		Deviation.	Tangent.	
I.	II.	III.	IV.	V.

VII.—*Blister Steel:* Average Weight, 982 grains.
[Tested under a Power of 51° 23'].

		°	°		°
B. H.	1	22·0	22·0	404	14·16
	2	22·50	35·6	703	15·54
	3	21·7	41·17	878	14·2
	4	21·4	45·27	1016	14·58
	5	23·4	49·7	1155	17·30
	6	22·34	51·23	1252	17·0
Mean		22·6	. . .	406	15·37

The power, after combination, in this last
series, No. VII., being determined under the
tension only of six plates, or a deviating power
of 51° 23, whereas the series Nos. V. and VI.
were tested under a deviating power of 71° 47';
the last column of No. VII., compared proportionally with the former, will be too high. But
the means of a more correct comparison is given
in a subsequent summary or abstract of these
experiments, aided by the application of an
uniform test.

SECT. II.—*Of the Magnetical Powers, in combination and separately, of Steel Plates (7·5
inches in length), of* SIMILAR DENOMINATION,
but of DIFFERENT QUALITIES *of the original iron.*

By this series of experiments, was intended
to be compared, the respective capabilities for

the magnetic condition, in various measures of combination, of steel of known but different qualities, in respect of the iron out of which it had been made.

The magnetical capabilities of *cast steel* from Swedish iron, marked *hoop*-L, and also of the *peculiar* cast steel from best qualities of foreign iron, being given in the last section, the results with the additional kinds examined, require only to be here stated.

No. of the Plates.	Maximum Power Separately.	Power in Combination.		Power after Combination.
		Deviation.	Tangent.	
I.	II.	V.	IV.	V.

VIII.—Cast Steel from best *Bradford Iron*: Average Weight, 979 grains.
[Tested under a Deviating Power of 49° 45].

Cb. H. 1	28·8	28·8	535	16·30
2	27·55	38·45	803	16·10
3	27·25	43·5	935	13·35
4	27·12	45·10	1006	14·45
5	27·22	47·0	1072	13·5
6	26·43	48·40	1137	15·35
Mean	27·49	. . .	520	14·57

IX.—The above re-hardened: the Numbers not correspondent in each Series.
[Tested under a Deviating Power of 50° 0'].

Cb. H. 1	26·53	26·53	507	15·5
2	27·52	38·55	807	16·32
3	28·5	43·8	937	16·16
4	27·55	46·3	1037	13·52
5	27·40	48·0	1111	16·0
6	28·30	50·0	1192	17·52
Mean	27·49	. . .	526	15·56

No. of the Plates.	Maximum Power Separately.	Power in Combination.		Power after Combination.
		Deviation.	Tangent.	
I.	II.	III.	IV.	V.

X.—Cast Steel from *commonest Foreign Iron*:
Average Weight, 943 grains.
[Tested at a Deviating Power of 50° 0'].

C cf.H.				
1	23·0	23·0	424	8·10
2	18·55	25·10	470	− 2·20
3	25·28	31·0	601	+ 18·0
4	25·12	34·30	687	16·35
5	17·12	32·0	625	− 6·30
6	25·30	37·15	760	+ 14·45
Mean	22·33	. . .	415	8·7

XI.—The above re-hardened: the Numbers in each
Series not certainly correspondent.
[Tested as the preceding].

C cf.H.				
1	17·10	17·10	309	− 4·0
2	20·32	20·30	374	+ 9·30
3	16·58	22·2	405	− 1·20
4	18·55	25·0	466	+ 3·28
5	20·25	26·50	506	9·10
6	19·12	28·42	547	6·25
Mean	18·52	. . .	342	3·52

XII.—Cast Steel from a *good quality of Foreign Iron*,
but not the best: Weight, 1027 grains.
[Tested as the preceding].

C f.H.				
1	16·30	16·30	296	11·38
2	18·43	29·5	556	14·52
3	19·37	38·10	786	16·35
4	17·34	43·6	936	12·58
5	17·30	47.18	1083	13·0
6	16·38	49·37	1176	11·46
Mean	17·45	. . .	320	13·28

With these several series of cast-steel plates, converted out of three different qualities of iron, are to be associated the series Nos. I., IV., V., and VI. of the foregoing section. The comparative magnetical qualities, separately and in combination, are referred to in the General Results; but it will scarcely escape observation, on the most hasty glance over the tables, what a remarkable superiority of magnetic capacity, in the *separate* plates, is exhibited in the steel from *Bradford iron*. In plates or bars of the *particular dimensions* here made use of—the steel from Bradford iron appears to be decidedly the best, for single plate magnets, which I have tried. But this superiority, as shewn by the result of combination, is evidently to be ascribed in considerable degree, though not entirely, to the insusceptibility of this steel for the *extreme* hardness of cast steel from the best qualities of foreign iron. The hardness, indeed, approximates that of shear steel from *hoop*-L iron. Still the capability of combining advantageously to a considerable extent constitutes a satisfactory quality in this steel—though inferior to that of the best description of cast steel.

The series x. and xi., representing the powers of steel plates from commonest foreign iron, are very unsatisfactory; the degree of hardness of the different plates being found to be most unequal. The re-hardening, it will be seen, injured, rather than improved, the original condition. This uncertainty of character, however, is not unimportant to have been determined. The determination of the powers in combination, in this case, is not at all satisfactory; because of the neutralizing influence of certain plates (Nos. 2 and 5 in series x., and 1 and 3 in series xi.), the effect of which was to diminish the power of the mass to a considerable, but an uncertain, extent.

A general notion of the comparative powers in combination of these twelve series may be obtained, by casting the eye in succession over the different tables. For a satisfactory comparison, however, of the magnetical qualities of the more characteristic kinds of steel, I have thrown into the form of a diagram (No. 5) the powers of the series No. ii., double shear steel; v., cast steel (hoop-L); vii., blister steel; and viii., cast steel from Bradford iron. So distinctly are the relative magnetical properties of these four kinds of steel here exhibited, in all the varieties of combination up to six plates of each kind, and an aggregate mass, in each, of near 6000 grains weight, as to characterize, I conceive, this method of determining and representing such varied powers as being at once elegant and conclusive.

Before, however, we can be prepared to draw *general* conclusions, as to the relative magnetical qualities, or degree of superiority of any of the various kinds of steel, under the different changes of condition which influence the results, two other elements in the investigation are necessary: viz. the effects of *alterations in the thickness*, or mass of the steel plates; and of *annealing*, or reducing the extreme hardness in the plates, whose magnetical properties have been given in this chapter.

The investigation of the effects of these two conditions, for certain practical objects of much importance, will now be described; postponing the propositional deductions to the end of a subsequent chapter, in order to embrace all the circumstances which seem to affect the General Results.

CHAPTER V.

THE Investigations of this chapter are in many
respects common to others, which have already
been described. Herein, however, a further,
and very important object was contemplated;—
the extension of facts bearing on the general
inquiry of this part of the work, and *with direct
reference to the improvement of sea-compasses.*
Thus the object becomes quite specific. This
object may be thus more particularly stated:—
*the determination of the best quality and denomi-
nation, and degree of hardness of steel, for each
of the different limited masses or combinations
requisite to be employed in the construction of the
directing magnets for sea-compasses.*

About 30 sets of steel plates were employed

in these investigations, comprising from 4 to 24 in each set; and a total number of near 300. They were of but two different lengths, 6 inches and 7·5 inches. Their breadths were from 0·42 inch to 0·75 inch. The weights of the 6 inch plates, corresponding very nearly in each set, varied from about 100 to 240 grains; and of the 7·5 inch plates, from about 230 to 430 grains. All but two or three sets were pierced with four small holes, for the convenience of combination as compass needles.

The degree of hardness comprised three general kinds, which were equable throughout their respective plates: viz., a spring temper—quite hard—and particularly annealed.

The *qualities* consisted of best cast steel, ordinary cast steel, cast steel of Stubs' peculiar manufacture, and cast steel from Bradford iron; and the *denominations* embraced those of "cast steel," "double shear steel," and an undescribed kind of steel (apparently cast steel), used at the Mint.

All these plates were magnetized in the same way—that is, to their utmost capabilities; and their powers were tried at the corresponding distance of two lengths of the plates, respectively, from the centre of the compass.

A selection of the experiments are here given, comprising the power of the plates separately;

their powers in combination; their reduced powers after combination, and, with several of the sets, their powers in combination when separated by thin discs of wood.

SECT. I.—*Experiments with Plates for Compass Needles of best cast steel, tempered equably throughout, at a spring temper.*

The particular degree of hardness of all the plates described under this section belongs to the specific kind distinguished in the diagram, Plate II., and in the table of characteristic letters (page 188), by the letter **T**.

Several series of plates of this description were employed of each of the two lengths of 6 and 7·5 inches. The 6 inch plates were of the uniform breadth, very nearly, of 0·42 inch, and the 7·5 inch plates of 0·5 inch. Their thicknesses, as shewn by the differences of weight, were dissimilar in the different sets.

It should be observed, that the order of the plates, when the powers were tried in combination in the two conditions of " contact," and " separated by discs of wood," was not exactly the same; that is, the plates were not always placed together in the same order of succession. The general results of combination, however, are not affected by this circumstance.

No. of the Plates.	Power of Plates Separately.		Power in Combination.	
	Maximum before Combination.	After Combination.	In Contact.	Separated by Discs of 0·125 inch.
I.	II.	III.	IV.	V.

XIII.—*Best Cast Steel:* 18 Plates of 6 inches:
Weight 97·5 grains each.
[Tested under a Deviating Power of 30° 58′.]

	o ′	o ′	o ′	o ′
C. T. 1	6·20	3·0	6·20	6.45
2	6·22	3·46	11·8	12.10
3	6·30	3·8	—	16·45
4	6·50	2·40	18·0	20·33
5	6·33	3·20	—	24·10
6	6·30	3·18	21·55	27·28
7	7·55	1·30	—	30·10
8	6·33	3·20	24·40	32·30
9	6·20	3·0	—	34·45
10	5·45	3·30	26·5	37·20
11	6·50	3·0	—	39·14
12	6·18	2·35	27·10	40·50
Mean	6·34	3·1	—	—

XIV.—*Best Cast Steel:* 23 Plates of 6 inches:
Weight, 143·5 grains.
[Tested under a Deviating Power of 36° 58′.]

C. T. 1	10·46	3·50	10·46	10·46
2	10·50	4·30	17·40	17·59
3	10·55	5·47	—	23·15
4	10·20	2·45	23·50	27·15
5	10·22	4·43	—	30·10
6	10·35	2·56	26·30	33·18
7	9·50	3·43	—	35·30
8	10·35	3·10	28·20	37·52
9	10·25	5·27	—	39·48
10	10·12	4·12	30·22	42·30
11	10·30	3·50	—	44·0
12	10·35	4·16	30·45	45·55
18	—	—	—	53·24
20	—	—	34·33	55·15
Mean of 20	10·29	3·56	—	—

No. of the Plates.	Power of Plates Separately.		Power in Combination.	
	Maximum before Combination.	After Combination.	In Contact.	Separated by Discs of 0·125 inch.
I.	II.	III.	IV.	V.

XV.—*Best Cast Steel:* 22 Plates of 7·5 inches: Weight, 231 grains.

[Tested under a Deviating Power of 32° 0′.]

	° ′	° ′	° ′	° ′
C. T. 1	10·15	4·58	10·15	10·15
2	9·45	4·30	16·8	17·30
3	10·22	4·30	19·40	22·30
4	9·45	3·58	21·12	25·22
5	10·4	5·15	—	—
6	10·10	2·45	23·10	30·10
7	9·50	4·40	—	—
8	9·35	1·45	24·50	34·20
9	9·30	4·38	—	—
10	9·50	5·55	26·10	38·10
11	10·15	4·52	—	—
12	9·23	2·25	26·50	40·8
22	—	—	32·0	51·53
Mean of 22	9·54	4·3	—	—

XVI.—*Best Cast Steel:* 19 Plates of 7·5 inches: Weight, 299·5 grains.

[Tested under a Deviating Power of 34° 10′.]

C. T. 1	9·26	5·45	9·26	—
2	9·20	3·50	15·0	—
3	8·58	2·17	18·15	—
4	8·55	0·45	19·32	—
5	9·15	4·30	21·30	—
6	9·22	5·20	23·10	—
12	—	—	28·45	—
19	—	—	34·10	—
Mean of 19	9·26	4.6	—	—

These four tables exhibit the magnetical qualities of *tempered* steel plates, designed for compass needles, constructed in the year 1838. The comparative powers, as combined in *contact*, of those of Table xv., are represented in the diagram, No. 6., in the curve marked as— " Cast steel tempered : *C. T.* 231 grains." The same series of plates, when *separated* by thin discs of wood (0·125 inch in thickness), is found, as in the case of former experiments, to obtain a rapid accession of energy, especially in the larger measures of combination, in the proportion, with 22 plates together, of the tangents of 32° 0′ and 51° 53′, or of 625 to 1275 ; shewing, ultimately, a doubling of the power by the spacing of the 22 plates. The relation of these powers, as augmented by the separation of the plates to the extent of 13 in number, to those of the same plates in contact, as well as to plates of several other different kinds, is shewn in the diagram No. 6, by the faint line-curve marked with dots.

Tables xiii. and xiv., in like manner, shew the effect of spacing the plates in the progress of combination; but there is no peculiar feature in them that needs to be noticed.

SECT. II.—*Experiments with Plates for Compass Needles, of superior qualities of steel of various kinds and denominations, made quite hard.*

The degree of hardness of the plates here referred to, is very nearly that heretofore indicated by the characteristic letter H. It was in all cases the result of a sudden cooling in a bath of salt and water (except in the case of steel from Bradford iron, which, when plunged into water being found to fracture, was cooled in oil), when the plates had been heated to a cherry-red —no further reduction of the hardness being suffered, than was absolutely necessary, after the manner of the treatment of files, for bringing the plates into form   As this process was accomplished by a first-rate file-maker, the measure of reduction was found to be very inconsiderable.

All the plates employed for the investigations of this section were manufactured at the works of Mr. Peter Stubs, at Warrington; and, for the most part, were pierced for combination in the usual manner. They were of somewhat different forms, as hereafter described. The sizes were correspondent in length to those of the sets used in the preceding section; but the breadth of each class was greater—those of 6 inches long being 0·56 inch broad, and those of

7·5 inches long, 0·75 inch broad. The former were of the average weights, in the respective series, of 119, 123, 136, and 242 grains; the latter of 393, 400, 420, 424, and 427 grains.

These various plates, which were constructed in sets of 4 to 8 of each description, comprised five different qualities or kinds of steel, distinguished in the tables by their characteristic letters—*CS.*, *SS.*, *CL.*, *M.*, and *C b.*, or cast steel from best Bradford iron.

1.—Experiments with five series of 6 inch compass plates, 0·56 inch broad.

No. of the Plates.	Power in Combination.		No. of the Plates.	Power in Combination.	
	Deviation.	Tangent.		Deviation.	Tangent.
XVII.—*Cast Steel: CS.* H. Weight, 123 grains each.			XVIII.—*Cast Steel: CL.* H. Average Weight, 136 grains.		
1	5·47	101	1	5·20	93
2	11·0	194	2	10·58	194
3	16·0	287	3	15·28	277
4	19·45	359	4	19·30	354
5	22·30	414	5	23·0	424
6	25·15	472	6	26·22	496
7	28·30	543	7	29·0	554
8	31 30	613	8	32·2	626
Mean	5·40	99	Mean	5·40	99

No. of the Plates.	Power in Combination.		No. of the Plates.	Power in Combination.	
	Deviation.	Tangent.		Deviation.	Tangent.

XIX.—*Shear Steel: SS.* H.
Average Weight, 119 grains.

XX.—*Mint Steel: M.* H.
Average Weight, 242 grains.

No. of the Plates.	Deviation.	Tangent.	No. of the Plates.	Deviation.	Tangent.
1	8·0	141	1	10·50	191
2	14·40	262	2	17·18	311
3	19·45	360	3	22·50	421
4	24·20	452	4	27·0	510
5	28·12	536	5	28·40	547
6	31·10	605	6	31·15	607
7	33·16	656			
8	35·22	710	Mean Power of Plates separately	10·10	179
Mean	7·40	135			

XXI.—*Cast Steel* from *Bradford Iron: C b.* H.
Average Weight of the Plates, 141 grains.

No. of the Plates.	Weight in Grains.	Maximum Power Separately.	Power in Combination.		Power after Testing.
			Deviation.	Tangent.	

[Tested with Test-bar H : Deviating Power, 34° 48'].

No. of the Plates.	Weight in Grains.	Maximum Power Separately.	Deviation.	Tangent.	Power after Testing.
1	147	10·35	10·35	187	7·10
2	144	10·28	17·56	324	7·10
3	139	9·54	23·31	435	7·15
4	135	10·3	27·41	525	7·23
Mean	141	10·15	—	181	7·14

2.—Experiments with five series of Compass Needle Plates of 7·5 inches in length, and 0·75 inch in breadth.

No. of the Plates.	Power in Combination.		No. of the Plates.	Power in Combination.	
	Deviation.	Tangent.		Deviation.	Tangent.
XXII.—*Cast Steel: CS.* H. Average Weight, 427 grains.			XXIII.--*Shear Steel: SS.*H. Average Weight, 400 grains.		
1	9·33	168	1	13·3	232
2	16·10	290	2	21·25	392
3	23·38	438	3	27·25	519
4	29·15	560	4	31·0	601
5	33·38	665	5	34·20	683
6	37·0	754	6	36·50	749
7	40·18	848	7	37·17	761
8	42·13	907	8	38·7	785
Mean	10·7	178	Mean	12·44	226
XXIV.—*Cast Steel: CL.*H. Average Weight, 424 grains.			XXV.—*Mint Steel: M.* H. Average Weight, 393 grains.		
1	9·5	160	1	8·12	144
2	17·0	306	2	15·40	280
3	23·30	435	3	21·30	394
4	28·20	539	4	27·15	515
5	33·20	658	5	32·49	645
6	36·55	751	6	36·15	733
Mean	9·14	163	Mean	8·52	156

XXVI.— *Cast Steel* from *Bradford Iron*: *C b.* H.
Average Weight of the Plates, 420 grains.

No. of the Plates.	Weight in Grains.	Maximum Power Separately.	Power in Combination.	
			Deviation.	Tangent.
1	405	13·8	13·8	233
2	435	14·0	21·32	395
3	412	13·26	26·40	502
4	430	13·28	30·0	577
Mean	420	13·30	—	240

The mean powers of the four series of plates,
xxii. to xxv., individually, as tested under a
combination having a deviating energy of 50°40',
were as follow:—Series xxii. 6° 34'; xxiii. about
neutral; xxiv. 7° 7'; and xxv. 9° 19'.

In the ten series of hard steel plates—that is
of plates of different kinds of steel, hardened
in a corresponding manner and at a similar
heat—we find the superiority, in simple magnetic capacity, of the steel from Bradford iron,
still preserved. But this advantage only extends to single or double plates; as beyond
double plates the best *shear steel* is found superior. In the five series of 7·5 inch plates, it may
be observed, that whilst the powers of the plates
of each kind, singly, are very dissimilar, the inequality gradually diminishes in combination;
and that on a combination of six plates of each
of the four series, commencing with the twenty-
second (the total weight of such combination
being about 2400 grains), a remarkable coinci-
dence in the powers of all the kinds occurs:—the
powers of six plates of the four series referred
to, being respectively, 37° 0', 36° 50', 36° 55',
and 36° 15'. Beyond this extent of combination,
the *cast* steel, according to the general law,
takes the lead, and ever afterwards retains its
pre-eminence.

The magnetical powers of the principal kinds

of steel, as exhibited in the compass-needle plates of 7·5 inches in length, are represented in the diagram, No. 6. The scale on which this diagram is constructed is just treble that employed in the previous one, No. 5; but as both the scales of "tangents" and "weights" are in the same ratio, the analogy of the diagrams is correctly preserved. The powers of the 6 inch compass plates were, in like manner, projected in a diagram; but the general character of the curves was so very similar in the two diagrams, that it did not appear worth while to engrave both of them.

As the effect of reducing the hardness of these various kinds of steel, was still necessary to be determined, before any general conclusions as to the best steel and temper for compass needles could be satisfactorily drawn,—I have reserved the propositional results until the completion of the Investigations on all the conditions requisite to be examined, which have been already noted, as affecting the magnetical powers of steel.

CHAPTER VI.

ON THE EFFECTS OF THE *ANNEALING* OF HARD STEEL
PLATES OF DIFFERENT KINDS AND MASSES, ON THEIR
MAGNETICAL QUALITIES, BOTH SINGLY AND IN PAIRS.

———

THE term *annealing* I here employ, in the sense
in which it is ordinarily used in the arts, to
designate the process by which brittle substances
are, by the peculiar application of heat, reduced
in their hardness, and rendered less liable to
break. The actual result of the process, indeed,
in the case of magnets or other articles made of
steel, is the same as *tempering;* and I merely
employ another term to indicate a process different from, or a result gained in a more satisfactory manner by such process than by, that in
ordinary use among smiths and cutlers.

In the Investigations of Part i. chapter v., the
general effects of various degrees of hardness
on the magnetical capabilities of *best cast steel*,
were determined, together with the modifications
resulting from varieties of size and mass. The

results thus obtained, however, only afforded
the general laws of that particular denomination—"cast steel," with the special powers, in
respect to the leading conditions by which the
magnetic energy is affected, in the same.

For the determination of the best quality,
denomination, and degree of hardness of the
magnets designed for sea-compasses—an inquiry
of great, indeed of national importance—further investigations became necessary. All that
seemed needful to be done in respect to *quality*
and *denomination*, indeed, has been brought forward in the last chapter. But inasmuch as great
differences are found to exist in the capacities
of perfectly hard plates, suitable for compass
needles, in steel of different *denominations*, especially betwixt the powers of *cast* steel and *shear*
steel; and as the capacity of hard *cast steel* is
found, when in single thin plates, to be very
inferior to that of either tempered cast steel, or
unreduced shear steel,—it became desirable so to
determine the effects of annealing, that both the
best condition for compass needles, in the several
denominations, might be ascertained, and some
fixed mode and standard of reduction might be
obtained for practical use.

The condition, as to temper, of the bars and
plates heretofore employed in these Magnetical
Investigations, was, with but comparatively few

exceptions, that into which the bars and plates, respectively, had been brought by the artizan or manufacturer who supplied them. But it was found that, in many cases, the plates or bars of any particular set, *ordered* to be of a certain, technically - understood, temper, and *intended* by the manufacturer to be of such temper uniformly, were of very different degrees of hardness; and that the temper of different sets, expected to be alike, was also very dissimilar.

Hence it became obviously desirable, for obtaining the most effective practical results, that some mode of tempering should be determined by which a more perfect measure and accordance in hardness might be obtained. In order to this, certain conditions, which seemed to be reasonably attainable, at least within definable limits as to size and mass in the plates or bars of steel employed, would be practically necessary. If the steel plates were to be reduced, or "let down," *after* being uniformly hardened to their utmost capability, then the condition would be—to obtain the equal reduction of the hardness by an exposure to a determinate and equable degree of heat: and if the effect were to be obtained by the simple and direct process of hardening to any particular degree from the state attained in being forged,—then all the plates must be exposed to an uniform and specific degree of heat, and suddenly cooled in the same precise manner, and with corresponding rapidity.

This latter process is the one recommended in many of our books, treating of the construction of artificial magnets. Dr. Gowan Knight, M. Coulomb, and others, were wont to employ, for some of their magnets, this process, tempering or hardening steel bars, by the first intention, "at a cherry-red heat."*

An important fact, however, in the phenomena of tempering, shews that this method will be very uncertain, and unequal in its results: viz. that the resulting degree of hardness depends, not on the heat to which the steel is raised only, but on the degree of heat taken in connexion with the rapidity of cooling. Hence very small steel wires, or extremely thin plates, may be brought to a suitable temper for drills or springs by being heated to a blood red and then cooled in a current of air; whilst the same process would produce in masses of considerable thickness, a very slight change from the original soft state. And, in like manner, thin plates or needles, plunged into cold water when heated to a blood red colour, will be found as brittle as glass; whilst thick bars, treated in the same way, will not be hardened, perhaps, beyond the ordinary degree of cutting instruments.

* Brewster's Treatise on Magnetism, pp. 284, 292, etc.

For the trial of this method, I employed several six-inch bars of cast steel of different thicknesses. Some of the bars, half an inch square, when only heated to this degree of redness (blood or cherry red), and then plunged into cold water containing some salt, were found to be of a very low temper, so that they could be easily filed; the effect, however, was by no means uniform. Other bars of the same length and breadth, but only one-tenth of an inch in thickness, became, by the like treatment, considerably harder. One of these thinner bars, for instance, weighing 634 grains, being heated in an open parlour fire to a low red, not near to cherry red, and then cooled in spring water, was found to have a maximum capacity for magnetism, equal to a deviation at two lengths from the compass of 19° 10', which fell, on the application of the test-bar H, to 5° 50'. The same bar, heated in a stove to full cherry red and cooled as before, had its magnetical capacity increased to 22° 30' deviating power, with a power of 14° 28' after the action of the test-bar. From the guidance afforded by the table at page 57, it appears, that the degree of hardness, in the first instance, was greatly below that of spring temper, T; whilst the temper obtained, in the second instance, was somewhat harder than that of ordinary cutting instruments,—about the same as would take place in hard cast steel, if "let down" by being heated, until the progress of oxidation had produced a straw colour on the surface.

This mode of tempering—varying in the result with the thickness of steel subjected to it, and rendered further variable, even in plates or bars of similar dimensions, by reason of the uncertainty of the degree of heat as determined merely by the hue of the incandescence—could not therefore be employed with any satisfaction for the purpose contemplated; viz. for obtaining an uniformity in the degree of hardness of different plates or bars of steel.

Hence some method, analogous in its steps to that employed by cutlers and smiths—the steps being first to harden to a state of brittleness, and then to reduce by the action of a moderate specific degree of heat—afforded the best prospect of attaining the desired object. In order, however, to this method being made effective, a tolerably uniform degree of extreme hardness would be requisite to be obtained as the first condition in the process. And this condition, especially in masses not greater than those employed in the investigations of the last two chapters, is found to be much more easily attainable than uniformity of a lower temper—because very great differences, within given limits, in the heat at which the plates are plunged into

water, occasion but comparatively little variation
in the resulting hardness.

The practicability of producing the required
uniformity, in extreme hardness, in the corresponding plates of any particular set constructed
out of the same bar of steel, is shewn by the
results of the experiments on hard plates of cast
steel and shear steel, in chapter IV.; especially
in the sets V. VI. VIII. and IX.—in which the
general equality of the powers of all the plates
of each set, respectively, both as to their separate
capacity, and as to their powers after combination, shews the satisfactory equableness of their
degree of hardness.

This principle, therefore,—the *annealing* of
bars or plates, first made as hard as possible,
by subjecting them to the same exact degree of
heat,—promised to afford something like the
uniformity of temper required to be produced.
And if this should be accomplished, then might
we determine with the greatest precision, the
relative magnetical qualities of any particular
temper, or degree of reduction of the extreme
hardness; and at the same time we should secure
a method whereby such precise temper might,
in the use of like kinds of steel, in all cases be
obtained.

This investigation embraced several points of

inquiry essential to the attainment of satisfactory results:—such as, Whether there might be any difference, in effect, betwixt a momentary application of heat to the hard plates, and the continuation of the same heat for a considerable period? Whether any difference might arise from the method of cooling the plates at the conclusion of the annealing process? What might be the relative effects of annealing, at any given heat, on the various kinds and denominations of steel? What might be the relation betwixt the degree of heat employed and the measure of reduction in the hardness? What might be the effect, on the same bars, of repeating the annealing process? And finally, whether hard bars of different *forms*—such as the straight bar and horseshoe forms—are similarly affected, as to their magnetical qualities, by annealing'.

Each of these points of inquiry was made, in some measure or degree, the subject of particular investigation by experiment.

The modes which occurred to me for the application of determinate and equable degrees of heat to the hard bars or plates to be annealed, proved to be the same as had been previously employed by Mr. Nicholson and. Mr. Stodart. Supposing that a higher degree of heat would be requisite for the reduction of hard plates to a spring temper, than what really proved to be needful, I commenced with a bath of melted lead in a vessel constructed for the purpose. The rapid change of colour of the plates, however, when laid upon the lead, just brought to a state of fluidity, shewed that the heat thus applied was already at a maximum. At the same time the method was not a very convenient one, as the steel was liable to become partially plated with lead, unless very carefully protected.

From lead I proceeded to a bath of linseed oil raised to a high temperature. Not having, in the first trials, a thermometer suitable for the determination of the progressive rise of temperature, I was obliged, that the results might be satisfactory, to proceed to the heat of boiling oil.* The effect, as will be seen, was, on some descriptions of plates, much greater than was required; but the effect produced, was still sufficient to guide to certain important practical results.

Finding, however, that this subject was of essential importance for the completion of the series of investigations in progress, I subsequently repeated the experiments on annealing at various degrees of heat from 300° to 550° inclusive, as indicated by the thermometer. The particulars of each class of experiments are annexed.

* The temperature at which linseed oil, and other fixed oils boil, is stated to be about that of 600°.

SECT. I.—*On the Effects of Annealing hard Steel Plates in boiling Linseed Oil.*

Two series of experiments, under this head, will be sufficient to be described—embracing, as these do, all the points which have been satisfactorily determined.

1. The first series consisted of twenty-four plates of seven different kinds as to denomination, quality, or dimensions. They were chiefly thin plates prepared for the construction of compass needles. All of them had been hardened originally to the utmost degree; but some had been subjected to a very slight reduction in putting them into form, after the manner of the treatment of files. These plates being tied up in two nearly corresponding bundles, comprising equal portions of the several kinds of plates, were placed in a kettle of pure linseed oil, in the progress of being heated. A few minutes after the oil was brought to the boiling point, which required a considerable time, the two bundles were removed, and one of them was laid on the flags to *cool gradually* in the air, and the other was *plunged into cold water.*

[Table XXVII.

Table XXVII.—Effects of Annealing at the heat of boiling Oil (A B), on various descriptions of hard Steel Plates, H., designed for Compass Needles.

Description of Plates.				Powers when hard.			Powers after annealing.		
No. of the Set.	Characteristic Letters.	Length in Inches.	Weight in Grains.	Maximum Single.	Deviation in Pairs.	Tested by Bars H or D.	Maximum Single.	Deviation in Pairs	Tested by Bars H or D.
I.	II.	III.	IV.	V.	VI.	VII.	VIII.	IX.	X.
				o '	o '	o '	o '	o '	o '
XVII.	CS.	6·0	123	5·40		4·37	9 3		2·24
				5·58	11·4	5·1	8·37	15·7	1·56
XVIII.	CL.	6·0	136	4·13		3·30	8·38*		3·48
				5·0	10·12·	3·42	9·22*	15·32	3·53
				5·0		3·53	8·44		3·45
				5·37	10·8	4·57	9·9	15·18	3·25
XIX.	SS.	6·0	119	7·27		5·32	9·51*		−4·18
				7·2	13·14	5·37	9 42*	15·13	−4 30
				7·16		6·13	9·28		−4·3
				7·20	14·40·	6·10	9·29	14·43	−4·12
XXII.	CS.	7·5	427	9·25		8·42	14·12*		+10·42
				9 20	16·10·	8·36	14·17*	21·28	10·37
				9·28		8·44	14·2		10·50
				9·33	17·22	8·53	14·4	21·23	10·43
XXIII.	SS.	7·5	400	12·7		9·45	13·30*		4·14
				12·30	21·25·	—	13·52*	18 6	4·32
				12·16		10·30	13 37		4·47
				12 30	20·45	10 55	14·4	18·6	4·51
XXIV.	CL.	7·5	424	8·44		8·14	14·3 *		11·6
				9·23	16·39	8·40	14·23*	21 33	11·5
				9·50		9·15	13·48		11·0
				8·43	17·0·	8·12	14·23	21·25	11·15
IV.	C.	7·5	590	15·25		14·20	18·20*		14·32
				15·32	26·0	14 25	18·30		14·47

The experiments on the effects of annealing are here thrown into a tabular form, as the most convenient, with a reference in col. i., by Roman numerals, to the set of plates in the foregoing tables, to which each of these plates respectively belongs. Those suddenly cooled, when taken out of the boiling oil, are distinguished in the table by an asterisk in col. VIII. The powers of the plates *in pairs*, as shewn in cols. VI. and IX., were determined when the plates of each pair were in even contact. Some of the powers of *pairs* in col. VI. were supplied from the former tables, and are here distinguished by a dot after the figures. All the powers here registered are those of the influence of the magnetized plates on the needle of a compass at the distance of twice the length of the plates respectively.

In addition to the particulars herein notified, in respect to the magnetical powers of the plates under the different circumstances referred to—a further inquiry was carried on at the same time: viz. as to the effect of the annealing process on the degree of magnetic energy possessed by each plate when first placed in the oil: in other words, as to the quantity of energy lost by the operation? The result of this inquiry is noticed in a future place.

The only observation, in anticipation of the summary of General Results, which occurs to me as being called for at this position of the Investigations on the Effects of Annealing, is the striking and important differences in the effects of the process on the two denominations of *cast* and *shear* steel. In the limited measure

of combination here employed—extending only
to pairs of plates,—and in the general thinness
of the plates subjected to the process, not exceeding, except in the last two, the thirtieth of
an inch,—the *cast* steel in every case *gained* in
magnetic capacity by the annealing; but the
shear steel derived comparatively small advantage, even in the individual or separate capacity
of the plates, whilst in combination there was a
loss of capacity, even in a pair of 7·5 inch plates
weighing together only 800 grains. In larger
numbers in combination, the annealing in boiling oil was found to be every way injurious
for magnetical purposes. The effects may be
perceived by mere inspection of two of the dotted
curves in the diagram No. 6—representing the
powers of two kinds of annealed plates—shear
steel of 400 grains, and cast steel of 424 grains,
each, and in combinations of four and six plates.

2. The results obtained by a second trial of
the effects of the annealing of hard steel plates
in boiling linseed oil, are shewn in the following
table. The experiment in this case extended to
a still greater variety of plates or bars (all being
originally of the hardness designated by the characteristic letter H); but the mean powers only,
of each variety, are here given. An additional
column, No. XI., shews the number of plates of

each description employed in this investigation; and though two columns are omitted from the series included in Table XXVII. yet the numbers of the similar columns are made correspondent in these two tables.

Table XXVIII.—Effects of Annealing on various kinds of hard Steel Plates, H.

Description of Plates.				Powers when hard, H.		Powers after Annealing, A B.		Amount of Plates used.
No. of the Set.	Charact. Letters.	Length.	Weight.	Mean Single.	Mean in Pairs.	Mean Single.	Mean in Pairs.	
I.	II.	III.	IV.	V.	VI.	VIII.	IX.	XI.

The following 5 Sets were immersed for two minutes in boiling Oil, and then plunged into cold Oil.

				° ′	° ′	° ′	° ′	
XVII.	CS.	6·0	123	5·56	10·55	8·33	15·6	4
XVIII.	CL.	6·0	136	5·28	10·13	9·19	16·9	4
XX.	M.	6·0	242	10·23	17·18	12·42	20·10	2
XXII.	CS.	7·5	427	9·4	16·10	13·51	21·47	2
XXIV.	CL.	7·5	424	9·15	17·0	14·8	21·53	2

The following Plates were 30 minutes in boiling Oil, and gradually cooled in the Air.

XXII.	CS.	7·5	427	10·16	17·50	14·27	22·1	4
XXIV.	CL.	7.5	424	8·54	16·20	14·31	22·25	2
II.	SS.	7·5	912	23·40	35·20*	16·8	} 19·50	2
I.	CS.	7·5	1000	22·34	33·32*	20·38		
VIII.	Cb.	7·5	1005	27·46	38·45	17·38	20·20	2
X.	C cf.	7·5	940	24·15	25·30*	15·12	17·40	2
XXV.	M.	7·5	394	9·38	15·40*	14·8	—	1
—	C.*	7·5	694	14·15	25·15	19·16	25·10	2
—	B.	7·5	640	17·24	—	15·6	—	1

These plates were taken from a set not included in the varieties of the foregoing tables. The set consisted of plates constructed for compound compass-needles, made of cast steel, and of the hardness approximating the degree H.

The general correspondency of the conclusions derived from this experiment, with those of the foregoing table (No. xxvii.), renders any particular observations, in this place, unnecessary. The principles and deductions flowing from the investigations, will have their proper place in the General Results.

SECT. II.—*On the Effects of Annealing, on hard Steel Plates, in hot Oil, at various Degrees of Temperature.*

Hitherto these investigations, in regard to the magnetic powers of steel as affected by degrees of hardness, have proceeded, mainly, on a comparison of the several ordinary distinctions of temper familiar to cutlers and smiths, and herein indicated by the characteristic letters S. E. T. H., together with the preceding experiments on the annealing in boiling linseed oil, AB. corresponding very nearly, in the effect on cast-steel, with the ordinary spring temper, T.

But whilst under these specific conditions, the advantage, as to magnetic capacity, was found, on all trials with straight bars of steel, to be in favour of the greatest hardness, when either large masses or combinations of many plates or bars were to be employed; it by no means followed that no condition of hardness intermediate betwixt H and T (a large interval as

to the degrees of hardness), might not afford a
capacity for a higher measure of magnetic energy,
for less considerable masses or combinations.

My attention was, in the first instance, directed
to the case of straight-bar magnets; though I
certainly did not anticipate the difference that
was actually found to take place when the bars
were turned into the horse-shoe form.

Thus far my investigations on the subject of
annealing had proceeded, when a notice of the
results on the tempering of steel, obtained in
the course of the inquiries of Dr. Faraday and
Mr. Stodart, on the effects of various alloys, etc.
fell into my hands. From the effects of annealing, as shewn in the last two Tables xxvii. and
xxviii., I had previously inferred (immediately,
indeed, on the completion of the experiments),
that many of the plates had been reduced in
hardness *below* that of the best temper for
compass needles. The deductions of Mr. Stodart
confirmed this inference, and, at first sight,
appeared in considerable measure to have super-
seded the necessity for the further inquiry I had
contemplated; that is, as to the connexion betwixt the degree of heat applied to hard steel,
and the resulting temper, or the measure of
reduction, in the hardness.

A little consideration, however, made it evident that the subject was of so much importance

as to call for special investigation on the principles, and by the mode of testing, etc. already
so extensively pursued; without which, indeed,
the actual effects of various degrees of heat on
the magnetical properties of hard steel bars
could not be really known.

Having provided myself with a chemical
thermometer, graduated up to the boiling point
of mercury, with a stove expressly constructed
for the convenient heating of the oil, I proceeded
with this investigation, in accordance with the
methods, so far as herein applicable, which had
been adopted for the experiments (Sect. I.) on
the annealing in boiling linseed oil.

Six sets of plates and bars (all the plates being
7·5 inches in length) were subjected to this experiment on annealing at various temperatures.
These consisted of *shear steel* plates, belonging
to set No. III. in the foregoing tables; *cast steel*
plates of set No. I; cast-steel plates of a set
designed for compass-needles described at the foot
of Table XXVIII.; *shear-steel* plates for compasses,
of the set No. XXIII.; *cast-steel* plates of like
description, of the sets Nos. XXII. and XXIV.,
together with the bars of a new five-bar horse-
shoe magnet made by Messrs. Stubs, the results
obtained with which belong to the next section.
All these plates and bars were, at the commencement of the experiments, in their original
state of *extreme hardness.*

These being arranged in six different sets, so that they might conveniently be subjected to the various degrees of heat designed, were placed in succession in the vessel of oil whilst on the fire. As the thermometer rose to each of the required temperatures of 300°, 350°, 400°, 450°, 500°, and 550°, one of the sets was removed and plunged into cold oil. The effect of the temperature of boiling oil having been before investigated, was not now repeated, but the results supplied, chiefly, from former experiments.

Previous to the annealing of the various plates and bars, their maximum magnetical powers were determined, both singly and in pairs, by the deviations at two lengths from the compass; and their strength, or tenaciousness of the magnetic condition, was likewise ascertained. After the annealing, the same qualities were again determined. I regret that the mode of testing before and after the annealing, was not the same. In the former case, the test was derived from a miscellaneous combination of the plates themselves, exhibiting, in the mass, the required energy; but this test, where the plates or bars are of various degrees of tenaciousness, as must be the case where shear-steel and cast-steel are combined, was found not to be very consistent or decisive. But in the latter instance (the test-

ing *after* the annealing) the results were obtained
by placing each plate in succession betwixt two
bundles of hard cast-steel plates of the required
energy, and these results were found quite satisfactory. Tables xxix. to xxxiii. comprise the
whole of the particulars, in regard to the several
series of plates, which seem to be necessary for
the object contemplated. All the plates, it
should be remembered, were of the same length,
viz. 7·5 inches.

Maximum Powers before Annealing.			Powers after Annealing.			
Deviation Single.	Deviation in Pairs.	Tested under Power of 50°.	Temperature of Oil.	Deviation Single.	Deviation in Pairs.	Tested under Power of 50°.
I.	II.	III.	IV.	V.	VI.	VII.

XXIX.—*Shear Steel:* *S* H. [Table iii.] Average Weight, 623 Grains.

° ′	° ′	° ′	°	° ′	° ′	° ′
17·36*	25·40	5·3	300	18·4*	26·3	— 1·10
17·24*	28·18	9·6	350	17·12*	27·20	+ 2·0
17·23	} 26·38	3·27	400	17·12	} 24·34	— 1·5
18·13		4·30	450	18·5		— 6·50
15·33	} 24·57	4·40	500	15·33	} 20·33	— 8·45
15·58		2·52	550	15·28		—10·15
16·37*	. . .	3·44	Boiling	15·37*	. . .	— 9·36
16·58	26·23	4·46	450	16·44	24·37	— 5·6

XXX.—*Cast Steel:* *C* *S* [Table i.] Average, 977 Grains.

17·50	} 32·56	15·18	300	17·28	} 31·54	14·35
20·46		18·0	350	20·45		16·50
18·55	} 33·42	15·0	400	19·22	} 32 50	13·40
22·8		17·40	450	23·18		10·40
19·45	} 32·0	16·35	500	22.26	} 29·27	2·5
18·3		15·27	550	22·6		0·52
18·0*	. . .	15·22	Boiling	21·34*	. . .	0·37
19·21	. . .	16·12	450	21·0	. . .	8·18

* The deviations, with an *asterisk* annexed, are the means of a pair of
similar plates. Those with a *dot* annexed, were supplied from former
experiments, in one or two cases derived proximately from such experiments.

Powers before Annealing.			Powers after Annealing.			
Deviation Single.	Deviation in Pairs.	Tested under Power of 50°.	Temperature of the Oil.	Deviation Single.	Deviation in Pairs.	Tested under Power of 50°.
I.	II.	III.	IV.	V.	VI.	VII.

XXXI.—*Cast Steel*, Plates for Common Needles, *C* H.
Average Weight, 694 Grains.

Deviation Single.	Deviation in Pairs.	Tested under Power of 50°.	Temperature of the Oil.	Deviation Single.	Deviation in Pairs.	Tested under Power of 50°.
14·38	} 24·13	10·32	300	14·2	} 25·6	9·58
14·0		4·42	350	15·28		7·35
14·38	} 26·28	10·32	400	14·57	} 26·0	8·50
14·53		10·52	450	17·5		6·13
15·22	} 26·50	12·45	500	18·12	} 25·48	1·35
15·10		12·35	550	18·2		— 1·56
...	...	...	600B	19·31*	25·38	— 3·35
14·42	25·50	10·20	450	16·45	25·38	7·7

XXXII.—*Shear Steel*, 7·5 inch Plates for Compass Needles, *S.*
[Table xxiii.] Average Weight, 402 Grains.

Deviation Single.	Deviation in Pairs.	Tested under Power of 50°.	Temperature of the Oil.	Deviation Single.	Deviation in Pairs.	Tested under Power of 50°.
12·30	} 21·0·	—	300	12·26	} 21·42	4·2
12·40		—	350	12·50		4·0
12·17	} 20·45·	—	400	12·47	} 20·55	— 1·0
12·29		—	450	13·12		— 2·30
12·40	} 21·0·	—	500	12·48	} 18·47	— 5·48
12·30		—	550—	13·17		— 9·59
12·23*	20·45	—	600B	13·52	18·6	—11·11
12·30	20·52	—	450	13·2	19·52	— 3·12

XXXIII.—*Cast Steel*, 7·5 inch Plates *C* and *L.*
[Tables xxii. and xxiv.] Average Weight, 427 Grains.

Deviation Single.	Deviation in Pairs.	Tested under Power of 50°.	Temperature of the Oil.	Deviation Single.	Deviation in Pairs.	Tested under Power of 50°.
10·40	} 18·20·	6·52	300	10·16	} 18·6	7·0
10·40		7·25	350	10·0		6·45
9·5	} 17·6·	7·30	400	9·23	} 18·20	6·36
10·0		8·0	450	10·58		6·48
10·40	} 18·20·	7·25	500	12·4	} 21·36	2·56
10·40		6·52·	550—	13·28		—4·18
9·4*	17·0	6·50·	600B	14·11	21·30	—5·22
10·7	17·41	7·16	450	11·29	19·53	2·55

These results, I may venture to state, are, in an experimental and scientific view, very satisfactory and beautiful. The second plates in each of the Tables xxix., xxx., and xxxi., exhibit, indeed, a measure of discrepancy; but column iii., wherein the effect of the testing is represented, shews that in each of these cases the particular plate in question was of a peculiar degree of original hardness, being, in the first two cases, harder than the others, and in the third case very much softer. But altogether the series of changes in the magnetic capacity of the separate plates, and in their degree of tenacious-ness, resulting from change of hardness, is, in each set of plates, strikingly consistent and in-structive.

The *shear-steel medium* plates, it will be ob-served (Table xxix.), derived no advantage of any importance by annealing, even in the low-est degree, in their individual capacity, whilst in pairs, there was an actual deterioration in capacity, at temperatures above 300°, rapidly augmenting with the increase of the heat. The result with *thin* plates, (Table xxxii.) exhibited a very slight gain of individual energy, but at the expense of a formidable deterioration in fixidity, with an absolute loss of energy in pairs of plates when the heat extended beyond 420° to 450°. The *cast-steel medium* plates (Table xxx.), for the most part, gained a little by the annealing, up to the heat of boiling oil; but in pairs, they exhibited, in every case, a deterioration, increasing with the degree of temperature to which the plates had been subjected. In the *thin* cast-steel plates, designed for compass needles, we find, as was reasonably to be anticipated from the previous experiments, a general improvement, (that is, after the first two of the series, which, being supplied from other experiments, are not equally satisfactory) the improvement extending to the powers of pairs of plates, as well as of single plates, throughout the series of temperatures. In the case of the plates, however, of intermediate thickness (Table xxxi.), we find, in perfect consistency with all previous conclusions, the gain, by annealing, limited mainly to the magnetical capacity of the plates separately,—a decided loss being found to take place in all the combinations annealed at a higher temperature than 400°.

Sect. III. — *On the Effects of Annealing, at various Temperatures, on the Magnetic Powers of hard Steel Bars, single and compound, of the Horse-shoe form.*

It yet remained to be ascertained experimentally, the effect of annealing on the magnetic capacity of hard steel bars of the horse-shoe form. So far as my experiments had hitherto gone, powerful combinations of the horse-shoe form derived, like those of straight bars, a decided advantage from hardness. In the instance of the large magnet, described at page 171, the permanent lifting power was more than doubled (nearly trebled) when the several bars, which, in their original state, were quite soft except for three or four inches near the poles, were moderately hardened nearly throughout their extent. It seemed reasonable to expect, therefore, from the previously observed analogies, found to be consistent in every considerable combination of hard straight-bar magnets, that the same law would hold in respect to powerful combinations of the horse-shoe form. The result of actual experiment, however, occasioned me, I confess, much surprise. The law, as to the *general* advantage of hardening the bars throughout, and that to a considerable degree, was amply verified; but the effect of *extreme hardness* was essentially different from what had been anticipated.

Two beautiful compact magnets, constructed for me by Messrs. Stubs, one of five bars, the other of fifteen, in which the bars were of fine cast steel, hardened to the extent and brittleness of files, were employed for investigating the law as to the best kind of temper, or measure of hardness, for instruments of the horse-shoe form. The bars of the smaller magnet weigh 4080

grains each; altogether 2·914 lbs. They measure
from the extremity of the curve to the poles
six inches, an inch in breadth, and, when combined, an inch in thickness. They are finished
flat and even at the poles. Those of the large
magnet, weighing together 8 lbs., are arranged
differently, — the three centre-bars being the
longest, and constituting at their poles an even
surface, but the others receding, by steps, being
successively shorter by the sixth of an inch.
Thus the full length of the centre bars, from the
extremity of the curved end of the instrument
to that of the poles, is 5·8 inches, whilst those
the most remote from the centre, being the
shortest, measure 4·8 inches. The entire thickness
of the fifteen bars in combination is 2·9 inches.

Each magnet was furnished with conductors
of good soft iron, made at different times, of a
variety of forms and sizes, in order to the selection, for general use, of the best. The bars of
both were magnetized, likewise, in a variety
of ways—the method found to be the most
effective, generally, here failing,—and that com-
municating the greatest power adopted for the
trial of the instruments.

But, to my great disappointment, the bars
were found extremely unenergetic, and their
combined powers not equal to some of those
of the commonest construction! The average

sustaining power of the bars, separately, was
scarcely 3 lbs., or about five times their weight.
The five-bar magnet carried, when first magnetized, and before the conductor was removed,
21 lbs.; but after the contact had been broken
it would only sustain 13 to 14 lbs. The fifteen-
bar magnet received a sustaining power of 33
lbs., which fell to 26 lbs. after the breaking of
the contact. These powers, I considered very
unequal to what was to have been expected from
the number of the bars, the quality of steel, and
the mode of construction. The effect of reducing
the extreme hardness, therefore, became an
important inquiry.

I commenced with the five-bar magnet, and
having determined the maximum sustaining
power of each bar in its original state, by a
spring balance, I placed the several bars in
heated oil, subjecting each one to a different
temperature; viz., one to that of 300°, and
others to the several temperatures of 350°, 400°,
450°, and 500°. On their attainment of the
desired heat respectively, they were plunged
into cold oil. The sustaining powers were
again determined, and found to be improved,
especially in those that had been subjected to
the highest temperatures. The power of the
five bars, when combined together, was likewise ascertained, and found to be considerably

augmented. Their retentiveness, with a view to further comparison, was also proved by the degree of energy remaining in each after being separated.

The same bars were, in the next place, all subjected to an uniform temperature in oil heated to 480° of Fahrenheit's scale, and their powers separately, in combination, and after their being again taken asunder, determined as before. But as the powers did not prove to be uniform, and as the bar, which had been heated to 500°, was much more powerful than the rest, a further experiment was made by subjecting the whole to the same maximum temperature of about 500°. Still, however, the powers of the several bars were not so generally accordant as was to have been expected,—the first three in the series being found to be feebler than the remainder, as if they had not been heated fully to the required temperature. And this, doubtless, from a defect in the experiment, was the fact.

A fourth, and final, experiment was therefore made, by placing the bars again in the oil, and heating it up to 500° (or rather 505°, the temperature which the oil ultimately attained), and keeping up the temperature for about a quarter of an hour before the removal of the bars. Hence there was every reason to believe, as the results indicated, that the bars had acquired the same temperature as the oil.

The results of these several experiments are exhibited together in the following Table.

Table XXXIV.

No. of the Bar.	Maximum Powers in the hard state н.	Powers after Annealing.			Powers after Annealing.		
		Temperature of the Oil.	Maximum Power of each Bar.	Power as reduced by com bination.	Temperature of the Oil.	Maximum Power of each Bar.	Power as reduced by combination.
		First Annealing.			Second Annealing.		
	lbs.	°	lbs.	lbs.	°	lbs.	lbs.
1	2·5	300	3·2	2·0	480	5·9	2·9
2	2·5	350	3·4	1·9	480	6·4	3·9
3	3·5	400	4·4	2·1	480	5·8	3·6
4	2·3	450	6·0	4·0	480	8·0	4·5
5	3·5	500	8·5	6·0	480	9·2	5·9
Mean	2·9	400	5·1	3·2	480	7·0	4·2
		Third Annealing.			Fourth Annealing.		
1	2·5	500	7·2	Not	505	7·9	Not
2	2·5	500	7·4	tried.	505	8·6	tried.
3	3·5	500	6·8	—	505	7·3	—
4	2·3	500	8·0	—	505	9·7	—
5	3·5	500	8·7	—	505	8·7	—
Mean	2·9	500	7·6	—	505	8·4	—

It should be observed, however, that the sustaining powers of the several bars, as here represented, can only be considered as proxi-mate;—as this mode of trial of the power of magnets is liable to so many differences from the varieties of surface, size, and degree of softness and quality of the conductors; as also from the measure of accuracy with which the surfaces of the magnet and conductor fit each other, as well as from the direction of the strain by which the adhesion of the conductor is tried. But, in order to the obtaining of results as satisfactory as possible, two rules were observed :— 1st. To take that amount of tension, as the proper power of the magnet, which could be sustained after the contact of the conductor with the magnet had been at least once broken after any repetition of the magnetizing process: and 2dly, to take, as the maximum power, generally, the highest power that had been twice obtained.

The gradual advancement of the average attractive powers of this series of five horse-shoe bars, by annealing at successively higher temperatures, up to that of 505°, is very striking and conclusive. For here we find the mean power of each bar of the series which, when quite hard (H), was represented by the sustaining of a weight of 2·9 lbs., increased, when annealed at an average temperature of 400°, to 5·1 lbs.; at 480°, to 7·0 lbs.; at somewhat less than 500°, to 7·6 lbs.; and when annealed at 505°, increased to 8·4 lbs. The reasons for going no further in the reduction of hardness, we shall see immediately.

The total lifting power of this magnet, in its final condition, was 25 to 26 lbs., after the repeated removal of the conductor. On the first trial, after being re-magnetized, it sustained a weight of above 38 lbs., before the conductor separated.

Thus, by this measure of annealing, we find that the separate powers of the bars were trebled, and the permanent power of the magnet, as a whole, was doubled. The sustaining powers, in comparison of the weights of the magnets, were equally satisfactory;—the entire magnet being now found capable of supporting nearly nine times its own weight, permanently, and each of the bars, separately, about fifteen times their weight respectively.

Such was the satisfactory nature of the results progressively obtained in the experiments on the annealing of a five-bar magnet of the horse-shoe form,—that I was encouraged to proceed to treat, after a similar manner, the large magnet of fifteen bars. The first trial of the effect of annealing on this instrument was made along with the other in the third experiment. The result, however, was not conclusive. For the thermometer having risen rapidly, and the boiler having been removed from the fire, on the first indication of a heat of 500°, and the magnets immediately taken out, the full degree of temperature intended had not been acquired by the steel. Hence these fifteen bars, which, in their originally hard state, possessed a lifting power, on an average, of about 2·5 lbs., only increased, in their mean magnetical power, to about 5 lbs., or double their former energy,— whilst some of the bars of the smaller magnet (that of five-bars), annealed more fully, had their original power more than trebled.

Subjecting, therefore, these fifteen bars to the final annealing, where the temperature, which rose at a maximum to 505°, was kept up, betwixt that and 500°, for above a quarter of an hour, —the effect, as to the augmentation of the magnetic capacity of the bars, separately, was quite satisfactory. For now each bar, when fully magnetized, was capable of sustaining a tension, as measured by the spring-balance, of from 7 to 9 lbs. (in one instance a tension of 10 lbs. was borne),—the general average being about 7·5 lbs.

These bars, when combined as one magnet, were found to have derived a satisfactory augmentation of energy. The original permanent power of this magnet, when quite hard, was only adequate to the support of about 26 lbs.; but now, after numerous trials, and the repeated breaking of the contact of the conductor, it would readily sustain 45 to 50 lbs.; a weight, which, on one trial, was gradually augmented to 56·6 lbs.,

R 2

before the load fell, being seven times its own weight.*

The loss of power in this magnet, by the combination,—indicated by the difference betwixt the aggregate sustaining powers of the bars separately, and that of the bars combined as one magnet,—was just about one-half: the loss, by combination of the bars, in the five-bar magnet, was about three-eighths.

From the great quantity of power lost by combination in the large magnet, with the disadvantageous effect (apparently) of the last annealing of bar No. 5, Table xxxiv., I inferred that the large magnet had been annealed at a maximum, if not beyond a maximum, temperature, for its best efficiency.

Thus the existence of a remarkable anomaly

* The proportion of load to the weight, carried by the two magnets by Messrs. Stubs, after being annealed, as has been described, shews, I think, that they are *very superior* instruments. My large horse-shoe magnet of nine bars, weighing 22 lbs., will only retain a permanent sustaining power of betwixt three and four times its own weight; and a compound magnet of seven bars, by Dr. Schmidt, weighing 16 lbs., belonging to the Royal Institution, London, which was supposed, I believe, to be a superior instrument, now sustains but 28 lbs., not twice its own weight; originally, however, the sustaining power was, no doubt, very much greater, it having been long in use, and by no means carefully treated.

betwixt the effects of annealing on large combinations of *straight-bar* magnets, and on similar combinations of magnetic bars of the *horse-shoe form*, or, in other words, the existence of this singular difference, that extreme hardness, which is of so decided an advantage in the one case, should have a contrary effect in the other,—seemed to be satisfactorily established.

The cautious philosopher, however, it occurred to me, might here question, whether, in the quality or kind of steel made use of in the construction of these horse-shoe magnets, there might not possibly be some peculiarity not belonging to the steel employed in any of the great variety of experiments hitherto made, by which the benefit of extreme hardness, in considerable combinations of, or masses in, straight-bar magnets, had been determined?

A conclusive experiment was undertaken with reference to this question. I ordered from the manufacturers of the horse-shoe magnets (who also were the manufacturers of the steel of which they were composed), five straight bars, similar in all respects, except as to form, to those of the five-bar horse-shoe magnet. Fortunately I was enabled to procure them *exactly as I required*,—of the same quality of steel as that out of which the other magnets had been made, c it off a bar of the same size, rolled out of the same ingot, and hardened by the same workman, precisely in the same manner. Each of them measured twelve inches long, one inch broad, and three-sixteenths thick. These straight bars, not being polished, were a little heavier than the others,— the total weight of the five being 3·164 lbs., instead of 2·914 lbs., and the difference of the weight of the bars, separately, about 250 grains.

These five straight bars, whilst in their condition of extreme hardness, were magnetized by the pair of 15 inch busk-magnets of 192 plates, so as to have the maximum power developed of which they were susceptible. Tried separately, their action on the compass, at two lengths distance, produced a mean deviation of 17° 40′; and their total powers, when in combination, a deviation of 35° 42′.

These bars were then subjected to the annealing process, at a temperature of 505°, and kept about that temperature for 15 or 20 minutes. Their powers, when again magnetized, were found, exactly as had been expected, to be *increased* when tried *separately*, but *diminished* when tried in *combination* : their separate deviations being now, on an average, 19° 9′, and their united influence 34° 22′. The effect of the testing, on their respective powers after the combination, was not quite so conclusive, there being an *apparent* gain, on the average, of about two

per cent. But, then, the test was not in both
cases the same; the testing being the effect only
of the united powers of the set of bars, which, in
the latter case, being weaker than in the former,
might occasion the small apparent discrepancy
in this quality of the bars in their two conditions
of hardness.

Still, in regard to the real object of inquiry,
the experiments with the five straight bars, in
their two conditions of hardness, H and A, were,
I conceive, quite conclusive. For they were in
strict accordance with the law so often demonstrated in this work,—that extreme hardness, in
considerable combinations of magnetic plates or
bars, is the most favourable condition for the
obtaining of the highest degree of energy.

Hence this important result, as to the reality
of the foregoing conclusion, of a certain change
in the form of a magnet occasioning a complete
reversion of the conditions requisite for the highest
energy, is obtained. For comparing the results
of these two series of experiments, in which
combinations of magnets were employed in all
respects the same, except as to form,—we arrive
at the conclusion sought to be established: viz.,
That the extreme hardness which contributed
to the maximum power of the combination of
magnets in the straight-bar form, detracted from
the power of a corresponding combination in the

horse-shoe form; and that the same degree of
annealing which doubled the power of the series
in the latter form, actually diminished the energy
of the series in the other form.

The investigations of this and the two preceding chapters, as a whole, go far towards the satisfactory determination of the important inquiry
proposed at the outset of chapter v. —as to the
best quality, denomination and temper of steel,
for each of the different limited masses or combinations, requisite to be employed in the construction of directing magnets for sea-compasses?
They likewise yield results, of a very conclusive
description, which are applicable to magnetic
instruments and apparatus generally. But, it
will be observed, that the general conclusions
are of a conditional, rather than of a positive,
character. For the results now, as well as heretofore, obtained, are demonstrative of the fact,
that no *general* or *universal answer can be given
to the question—What is the best denomination
of steel, or what is the best temper or degree of
hardness for magnetical purposes?* For denomination, and mass, and hardness, and form, are,
severally, qualities of varying influence in their
magnetical relations—changing the comparative
resultant energy, as we have already seen, with
the changes in these qualities, according to the

peculiar laws of magnetism, and their mutual
influences, belonging to each of these conditions
respectively.

Thus each case, of a number, where there are
differences in quality or denomination of the
steel, of mass or extent of combination, of hardness or of form,—becomes, in certain measure, a
special case. But we have now attained to such
an extent of information on these various influencing qualities, and on the mode in which they
severally and unitedly operate on the magnetical
powers of bars or plates of steel, whether single
or compound,—as to be enabled, on the statement of any particular requirement, to determine
with much satisfactoriness and practical facility,
what the best denomination and quality of steel,
and what the best degree of hardness, for such
a magnet, actually are. For the various propositions, derivable from the foregoing investigations, which are here thrown into the usual form
of General Results—will be found, I expect, to
comprise the great majority of cases likely to
occur in Practical Magnetics; whilst principles
will be found developed, I hope, calculated to
afford, by close analogy, some useful measure of
guidance in all. In these Results, I may add,
will be found embodied certain additional facts
derived from the experiments of this chapter,
which do not altogether appear in the printed
tables.

RESULTS.

THE three preceding chapters, IV. V. and VI., comprise the examination of the magnetical properties of a considerable variety of specimens of steel of various qualities, of different denominations, and in the several conditions betwixt that of extreme hardness (H), and that produced by annealing in boiling linseed oil (AB). The measure and kind of knowledge hereby acquired, in respect to the phenomena of magnetism in plates and bars of steel, will be the most conveniently epitomised, compared and illustrated, in our usual method of results. Some of these investigations, indeed, are, in substance, similar to others already described; but not in all the circumstances connected with the experiments. For in each case, whether entirely new, or in a great measure corresponding with what has gone before, there will be found certain differences, the knowledge of the effects of which was requisite, in order to the determination of the nature and measure of importance of all the circumstances affecting the magnetic energy in permanent artificial magnets.

To facilitate the comparison of the effects of *denomination* and *quality* on the capacity for, and tenaciousness of, the magnetic condition in HARD STEEL [temper H], I here prefix a table comprising a new investigation made with a selection out of the foregoing varieties of steel, comprising all the more characteristic differences, with the comparative degrees of the retentiveness, as determined *by an uniform testing* of all the plates under the same exact measure of violence. Column I. will readily shew to what set of plates, and column II. to what class, each of the different plates belongs.

Table XXXV. — General Results concerning the Effects of *Denomination* and *Quality* on the capacity for, and tenaciousness of, the Magnetic Condition in straight Bars and Plates of Hard Steel,—Temper H.

Set of Plates.	Class of Steel Plates.	Weight in Grains.	Maximum Power.		After Testing under Power of 50° 40′.	
			Deviation.	Tangent.	Deviation.	Tangent.
I.	II.	III.	IV.	V.	VI.	VII.
I.	*CS.*	986	22·30	414	18·45	339
—	—	—	21·4	385	18·24	333
X.	*C.cf.*	943	25·28	476	0·15	4
—	—	—	25·12	471	0·48	14
VIII.	*C. b.*	979	27·25	519	17·23	313
—	—	—	27·12	514	14·22	256
V.	*CL.*	952	20·37	376	18·0	325
—	—	—	21·42	399	18·40	338
IV.	*CS.*	607	11·25	202	8·0	141
—	—	—	12·0	213	9·10	161
*	*C.*	700	12·50	228	10·44	190
*		—	15·5	270	12·38	224
XXV.	*CM.*	394	7·50	138	6·22	112
—	—	—	8·50	155	7·22	129
II.	*SS.*	890	23·37	437	13·13	235
—	—	—	21·15	389	13·0	231
III.	*SS.*	632	16·38	299	2·5	36
—	—	638	17·20	312	2·12	38
* *	*B.cf.*	640	16·26	295	12·38	224
* *	—	—	15·58	286	12·30	222

The Plates marked * column i. were from an undescribed quality of Cast Steel. Those marked ** of Blister Steel from commonest Foreign Iron.

The various results, derivable from the investigations of the three preceding chapters, may, with respect to *straight bars* and plates, be conveniently arranged under five classes; besides the *peculiar* results as to magnets of the horseshoe form. And it will be important to observe, that the results belong to the particular dimensions and masses of steel herein employed (the length being uniformly 7·5 inches or 6 inches), or to combinations of plates of like sizes. For facility of description and reference, I may distinguish three classes, as to mass or thickness:— *thin* plates, such as those designed for compound compass-needles, being of the description employed in Tables XIII. to XXV. ;—*medium* plates, such as those employed in the investigations of chapter IV.;—and *thick* plates or bars, being of a mass greater than the preceding. In 7·5 inch plates, the *thin* kind would extend up to about 500 grains weight; the *medium* from 500 to 1500; and the *thick* from 1500 grains upwards, In 6 inch plates about half these weights.

Were more accuracy needful, the proportion which the area of the end, or transverse section, of a bar, bears to the superficial area of the four sides, might afford a mode of discriminating, proportionally, the relative masses, whatever might be their lengths.

SECT. I.—*Results in respect to Difference of* DENOMINATION *in the* SAME QUALITY *of Hard Steel.*

In speaking of *hard steel* (H), or of "*steel quite hard*," it should be understood, that the highest degree of hardness produced by immersion, at a bright red heat, into a saturated solution of salt and water, is meant. The result of this process on cast-steel corresponds with the hardness of the best files.

———————

1. That the *magnetic capacity* differs in each *denomination* of hard steel, being (as shewn in comparatively small masses) the lowest in denominations susceptible of the greatest hardness.

By *magnetic capacity*, I here mean the measure of energy which any bar or plate is capable of *receiving*. Thin or medium bars or plates exhibit their relative capacities, in a considerable degree, by their sustaining as well as their recipient capabilities.

The capability of sustaining the tension of a high degree of magnetic energy, it has been already shewn, exists, in the case of large masses, under different conditions from those of this result,—where the ultimate energy in denominations of the lowest capacity for magnetism becomes the greatest, by virtue, not of the capacity, but of the superiority of such denominations in fixidity.

Within the limits, however, of the masses investigated in the three preceding chapters, the proposition above stated is generally borne out: in the case of the thinner plates the result is always found consistent.

2. That in thin and medium plates made *quite hard*, shear steel possesses a higher capa-

city, and exhibits a greater energy in the *individual plates*, than blister or cast steel, and cast steel the least of all.

In the comparison of hard *thin* plates, such as those of six inches in length of 100 to 200 grains weight, or of 7·5 inches of about 400 grains,—we find this result constantly maintained. Thus, in regard to the powers of *thin* 6 inch plates, of best quality, we find these mean results:

XVII. *Cast steel*, weight, 123 grains; mean power, $\overset{\circ}{5}\overset{\prime}{\cdot}40$
XIX. *Shear steel*,   "   119   "   "   7·40

And in 7·5 inch plates we find:—

XXII. *Cast steel*,   "   427   "   "   10·7
XXV. *Mint steel* (cast)   393   "   "   8·52
XXIII. *Shear steel*,   "   400   "   "   12·44

In the comparison of *medium* plates—such as those of 7·5 inches in length, weighing 800 to 1000 grains— we find the powers of the different denominations still supporting, though in a very inferior degree, the above proposition:—

Table I. *Cast steel,*   986 grains; mean power $\overset{\circ}{20}\cdot20$
II. *Double shear,* 890   "   "   22·45

Again, V. Cast steel, 952   "   "   20·27
VII Blister steel, 982   "   "   22·6

The curves belonging to these three denominations of steel, as represented in diagram No. 5, shew, that the superiority of hard shear-steel over hard cast-steel continues to a much larger extent of mass or combination than 1000 grains, or of that of any one of the plates which I employed.

3. That the comparative magnetic powers of different denominations of steel change their relation to each other *in combination;* each deno-

mination, under powerful combinations, exhibiting a degree of effectiveness according (apparently) to its susceptibility for *hardness.*

Thus we find that *shear* steel, which exhibits a much higher capacity for magnetism in thin or medium plates than *cast* steel, is very inferior when a number of plates are combined together as one magnet.

In the case of Messrs. Stubs' peculiar steel,—Tables I. and II. shew these differences very satisfactorily, betwixt the powers of cast and shear steel in combination.

Six plates of *cast steel,* of 986 grains each, exhibited in combination a deviating power of 53° 44′; whilst six plates of *double shear,* of higher magnetic capacity in the separate plates, had a power only of 49° 0′.

Again:—Six plates of *cast steel* (Tables V. VI. VII.) from hoop-L iron, of 952 grains weight each, evinced a power in combination of 55° 30′ in one set, and of 54° 20′ in another; whilst six plates of *blister steel,* of 982 grains each, exhibited a united deviating action of 51° 23′.

Referring to diagram No. 5, we observe that the advantage of the best double shear-steel over the best cast-steel, when both are *quite hard,* continues up to the combination of about three medium plates, or near 3000 grains weight in the total mass. Beyond this measure of combination, the cast-steel, from hoop-L iron, assumes its characteristic superiority.

That the difference betwixt the powers in combination of the cast and shear steel is mainly to be ascribed to their unequal indurative capabilities, seems strikingly to be indicated by their respective deterioration in the individual energy of the plates, by the effect of the same degree of tension or violence to which both qualities had been subjected. Both sets of plates had been subjected to a magnetic power indicated by 50° 40' of deviation (Table xxxv.); but this degree of violence, which had reduced the *cast*-steel plates (set I.) on an average from 21° 47' to 18° 35', produced in the plates of double shear-steel (set II.) a reduction of power from 22° 26' to 13° 6'. Tested under a tension of 71° 47', the same denominations fell in still greater proportions; the cast steel on an average from 20° 20' to 12° 6', and the shear steel from 22° 45' to 4° 58' (p. 190). In respect to the apparent difference of hardness of shear steel and cast steel, when both are hardened to the utmost, it must be admitted that the conclusion is derived rather from analogy, with reference to their respective magnetical properties, than from direct experiment. It is possible, therefore, that the magnetical differences may be owing to some other cause.

4. That *cast steel*, being capable (as it would seem) of the greatest hardness, is, as a denomination, the most effective in large straight-bar magnets, whether consisting of single massive bars, or of considerable combinations of plates or bars.

A general inspection of the Tables in the three foregoing chapters, and of the diagrams Nos. 5 and 6, will satisfactorily support this proposition as to the relative efficiency of cast and shear steel. The comparative efficiency of *blister* steel has not been sufficiently investigated, though so far as the trial of one set of plates may be relied on, its relative powers have been projected in the diagram No. 5.

It must be kept in mind, however, that we here speak of

s

the comparison of steels made from the highest qualities
and of the same description of foreign iron, for it will
be seen that cast steel from English iron possesses,
apparently, very inferior capabilities for hardness.

SECT. II. — *Results in respect to Difference of*
QUALITY *in the same* DENOMINATION *of hard
Steel.*

5. That the magnetical properties of the *same
denomination* of steel vary with the qualities of
the iron out of which the steel may be manufactured.

> This general proposition has been discussed in Part I. of
> this work, and constitutes the foundation principle on
> which the *quality* of steel is proposed to be determined
> by the magnetical test.

6. That in *cast steel* from various kinds of
iron, the steel from those of highest repute in
commerce, so far as yet tried, exhibit generally
the best magnetical properties.

> This result again corresponds with the views already
> submitted; and so far as my experiments have yet
> proceeded, have been satisfactorily verified. One ex-
> ception indeed, as to the value in the market of a
> particular "mark" of foreign steel, has been met with,
> but I have great reason to believe that that description
> of foreign iron is by no means duly estimated. It
> bears at present but a low price in the market, whilst
> it appears to possess first-rate qualities.
>
> The general inference deducible from this proposition is,
> that the purest iron smelted with charcoal, is the most
> valuable for the manufacture of steel.

7. That cast steel, quite hard, from the first qualities of foreign iron, appears to be the best for magnetical purposes, where straight bars are employed, and where considerable energy is required; but that for single thin or medium plates, the cast steel from Bradford iron exhibits first-rate capabilities.

This result is drawn from the experiments given in Tables I., V., and VIII. to XII. Taking the *combination* of each set of steel plates to the extent of six, we find that cast steel from the *commonest* foreign iron exhibited a total power only of 37° 15′ deviation in one case [Table X.], and of still less, (28° 42′), in another trial of the same plates after re-hardening [Table XI.] Cast steel, again, from a *good* quality of foreign iron, yielded, in six plates, a power of 49° 37′. The same denomination of steel from Bradford iron gave a power, in six plates, originally of 48° 40′ [Table VIII.], and after re-hardening of 50° 0′; whilst the cast steel from best Swedish hoop-L iron sustained the magnetic energy in six plates to the extent of occasioning a deviation in the compass-needle at two lengths distance of 55° 30′ in one set [Table V.], and of 54° 20′ in another set [Table VI.] The power yielded by Messrs. Stubs' peculiar steel was somewhat lower, being 53° 44′ [Table I.] The small differences in the weight of the plates have not been taken into account in these comparisons, not being considered important; but in the diagram No. 5, the cast steel from hoop-L iron and that from Bradford iron are fully compared.

In regard to *single plates* of thin or medium masses, the steel from Bradford iron exhibits, as shewn both in the Tables VIII. and IX., and in the diagram No. 5,

s 2

an observable superiority, which superiority is maintained up to the extent of a combination of about three plates, comprising a mass of 3000 grains or upward.

SECT. III.—*Results as to the Changes in Magnetic Properties produced in hard Steel (straight-bar or ruler form), of different Denominations and Qualities, by* ANNEALING (AB.); *that is, reducing the hardness, by subjecting the Plates to the heat of* BOILING *Linseed Oil.*

Some of the results under this head may appear to have been anticipated, in substance, by those at the end of chapter III.; but as the present deductions refer to a specific degree of reduction of hardness on various kinds of steel, their omission would have rendered the series of investigations incomplete.

———————

8. That the *annealing* of hard steel plates or bars (being the same thing, in effect, as tempering), changes the relative positions of the different denominations of steel as to their magnetical properties.

> The slightest examination of Tables XXVII. and XXVIII. will be sufficient to establish this proposition; whilst the inspection of diagram No. 6, will shew the progress of the magnetic changes produced by annealing, in the degree now under consideration, in two denominations of steel, *cast* and *shear*, under the action of successive additions of magnetized plates in combination.

9. That, in the case of *thin* plates, each kind and denomination of steel *gains* power, in the

individual capacity of *hard plates*, by annealing in boiling oil; whilst *medium* or *thick* plates, or combinations of plates, generally suffer loss by the same extent of reduction of temper.

The *thin* plates experimented on, consisted of 6 inch plates of 119 to 242 grains in weight, and of 7·5 inch plates of 394 to 590 grains. In every kind, amounting to nine varieties either of quality or denomination, the annealing of the plates occasioned an augmentation in the capacity for magnetism; the greatest augmentation being in the denomination of cast steel. See Tables xxvii. and xxviii.

In the case of *medium* plates, however, such as those of 7·5 inches, weighing 900 to 1000 grains, a *loss* of power was occasioned by such annealing of the hard plates in every one, out of six in number, whereon the trial was first made. The extent of loss varied, in the different kinds of steel, from about 2° to 10°.—Table xxviii.

In subsequent experiments on the annealing of *cast-steel* medium plates, one example of gain occurred to the extent of 3° 34′ deviation.

When the effect of annealing was examined on larger masses—such as a combination of those of *medium*, or even of the *thinner* plates—the loss of power by the reduction of hardness was still more striking. Thus *two* plates of 1005 grains each (cast steel from Bradford iron), were reduced in their combined deviating energy from 38° 45′ to 20° 20′, by being annealed. Table xxviii.—So also, in the case of the thin 6 inch plates, set No. xix., four plates of shear steel, in combination, had their powers reduced by annealing from 23° 20′ to 18° 40′.

10. That in *thin* single plates, *shear* steel gains *slightly* and *cast* steel *greatly*, by annealing in boiling oil; whilst in considerable combinations of plates, *shear* steel suffers much more deterioration in its magnetical powers than cast steel.

The proportional *gain* of cast steel over shear steel, in single plates, by *annealing*, is shewn in Table xxvii. Cast steel, in thin 6 inch plates, is there found to gain about 60 per cent., and in thin 7·5 inch plates about 60, separately; whilst shear steel of like dimensions, only gains about 24, and 10 per cent. in the separate plates. In *pairs* of plates, similar cast steel plates of 6 inches (being less in mass, proportionally, than the 7·5 inch plates), gain 40 to 50 per cent. by annealing, and the 7·5 inch plates 25 to 35 per cent.: whilst shear steel plates in pairs of like dimensions gain but from 1 to 15 per cent. in the 6 inch plates of 119 grains, and actually *lose* from 10 to 15 per cent. by annealing in the 7·5 inch plates of 400 grains.—Table xxvii.

In larger combinations than that of pairs—the cast steel continues to exhibit an advantage by annealing, up to a combination of about 7 to 8 plates; the curves crossing, in the case of the 7·5 inch plates, betwixt the seventh and eighth plate. In the shear steel, the curves, it will be observed, cross a little beyond the first plate.—Diagram No. 6.

11. That the superiority of shear steel over cast steel, as exhibited in the comparative capacity of the two kinds, in thin or medium plates, hardened in a similar way, is apparently owing

to its inferior hardness, which, in such small
masses, is favourable to capacity; but when the
cast steel is reduced by annealing, or tempering,
to a similar measure of tenaciousness of the
magnetic condition, or correspondency of hardness, (?) then the apparent superiority of the
shear steel mainly disappears.

In this proposition, the results given in Nos. 2, 4, and
some others, are generalized, and the effect of annealing
on cast steel added thereto.

12. That in the process of annealing, no
difference is found in the resulting magnetic
capacity of steel plates, whether, after being
subjected to the heat of boiling linseed oil, they
are cooled suddenly or slowly.

For the determination of this fact, which might have
been of much practical importance, a variety of experiments made on two different occasions were instituted,
which are briefly described in the text of the foregoing
chapters; the results have been reserved for this place.

In this investigation double sets of *thin* and *medium* plates
were employed, some of 6 and others of 7·5 inches in
length. On the first trial these consisted of six varieties,
and on the second of nine varieties, of the plates previously made use of in the experiments referred to in
the tables of the last three chapters. A careful examination of Table xxvii., in which is registered the first
series of experiments, will verify, I think very satisfactorily, the above proposition.

The powers of each set in pairs, when compared after
being annealed, shew, indeed, a very slight inferiority

in the plates cooled slowly (col. ix., Table xxvii.): but on comparing the resulting powers, after application to the test-bar (col. x.), we find an equivalent compensation in the superior endurance of the test, in the last four sets of plates. And in the second series of experiments (Table xxviii.), we find the rule still further confirmed, and the *apparent* objection just noticed further weakened, by the fact—that, in a comparison of the sets xxii. and xxiv., those plates which were cooled slowly, after being annealed, exhibited a somewhat larger capacity, both singly and in pairs, than those which were plunged into cold oil.

13. That the *measure of reduction of hardness* which takes place on the annealing of hard steel plates or bars, depends on the degree of heat, not on the method of cooling, to which the steel, during the process, is exposed.

This proposition we obtain, by simple inference, from the illustrative notes under the preceding one.—Whether, however, the degree of reduction in temper, exactly corresponds in plates or bars of different proportional masses, has not been determined. Some very thin plates, indeed, appeared to have lost more of their hardness by the annealing than the others, and more than was due to the degree of heat to which they had been subjected: but in this case the quality of the steel might have been injured in the process of hardening, or the original hardness might have been inferior in these, to that of the other plates with which they were compared.

14. That no essential difference was observed in the effect of the annealing process, whether

the hard steel was immersed in the boiling oil
only for a minute or two of time, or for as many
hours; provided, only, the steel had become
duly heated throughout its substance.

This proposition is founded on the examination of the
effects of annealing in sets XVII. XVIII. XXII. and XXIV.,
as exhibited in the two cases of Tables XXVII. and
XXVIII.; and also, in the comparison of sets XXII. and
XXIV. in the two sections of Table XXVIII. In the first
instance we have the case of the plates being put into
very hot oil, and remaining therein nearly two hours,
whilst it was gradually advancing to the boiling point,
in comparison with the exposure of plates of the same
set to the heat only for two minutes, and that whilst
the oil was boiling. In the second instance we have
similar plates, brought into comparison, when one part
or selection had been immersed for thirty minutes, the
other only two minutes in the boiling oil.

15. That when *magnetized* plates or bars are
subjected to the process of annealing, the effect
on their original energy appears to be somewhat
deteriorating,—slightly so in the case of best
cast steel, and considerably so in the case of
shear steel.

The impracticability of keeping the plates and bars free
from the influence of each other, when placed in the
same vessel, though adjusted generally in pairs with
diverse poles in contact, renders this result not very
conclusive.

In the case of *cast-steel* plates, of the sets XVII. XVIII.
XXII. XXIV., the average loss of power observed on the
completion of the process (the power being previously

reduced by testing, as shewn in col. vii., Table xxvii.),
was about seven per cent. But the shear-steel plates
(from sets xix. and xxiii., Table xxvii.) lost, on an
average, about 24·5 per cent. This large amount of
loss, however, might probably have been partly occasioned by the deficiency in the property of fixidity in
the annealed shear steel, whereby unfavourable contact
with the more powerful plates of cast steel would have
a highly deteriorating influence.

SECT. IV.—*Effects of* ANNEALING *on the Energy
of hard straight-bar or plate Magnets, at
degrees of Temperature* INFERIOR *to that of
Boiling Oil.*

16. That the annealing of thin and medium
plates, whether of *best* cast steel or shear steel,
at temperatures not exceeding 400°, seems to
produce no material change in their *individual*
magnetical capacity; though a sensible deterioration in their tenaciousness of the magnetic
condition begins to take place below such
temperature.

Comparing the deviations given in columns i. and v. in
Tables xxix. to xxxiii. inclusive, in the experiments
on annealing at the three temperatures of 300°, 350°,
and 400°, we find only a gain of about 1° 10′ in the
whole,—indicating a proportional augmentation, by the
annealing of these various plates, not exceeding a two-
hundredth part of the original magnetical capacity.

17. That, annealed at temperatures *above* 400°,
both medium and thin plates of *best cast-steel*,
gain, singly, in magnetical capacity; the thin

plates in an increasing ratio up to the boiling temperature, and the medium plates attaining a maximum, at the temperature of about 500°, but these medium plates falling at the boiling heat so as to approximate the capacity in the original hard state.

It is shewn, in Result No. 9, that medium plates of steel of the various qualities and denominations yet subjected to experiment, generally *lose* by the process of annealing at the boiling temperature. In the case of best cast-steel, however, in 7·5 inch plates of about 1000 grains, the resulting *capacity* for magnetism, after annealing at a boiling heat, does not appear to be materially different from that of the original hard state. In one experiment, Table xxviii., such a plate suffered a loss of nearly 2° by the annealing; and in another, Table xxx., there was a gain of 3° 34′, in the deviating power, after being subjected to the like treatment.

Cast steel from English iron, or common foreign iron, seems to follow the law of shear steel very nearly, having the highest capacity for magnetism, except in very thin plates, when quite hard.

18. That hard *shear steel*, annealed at temperatures *above* 400°, seems to gain, though very slightly, in capacity, in *single thin* plates, up to the boiling temperature; but loses very considerably in *pairs* of plates at the higher temperatures. In *medium* plates, any reduction of the extreme hardness (unless such as might result from the lowest temperatures) appears to deteriorate their capacity, singly, for magnetism, and very greatly so in combination.

The *gain* of power by annealing in *thin* shear-steel plates, when tried at various temperatures, was only, on an average, 0° 32′ (Table xxxii). The greatest gain, in any one instance, did not exceed 1° 30′, or about 12 per cent. The same *pair* of plates, however, in which this gain in their separate capacity was exhibited, were found to have *lost* a still greater proportion of their united powers, when in combination.

In the case of shear-steel plates, betwixt the thin and medium weights (Table xxix.), there was a general loss, on the average, in the powers of the single plates, by annealing at all the different temperatures. In a shear-steel medium-plate of 912 grains (Table xxviii.) the loss by annealing was much more considerable.

19. That the annealing of any plates or bars of a straight or ruler form (designed for magnetical instruments), whatever be their denomination or quality, is always injurious to their *tenaciousness* of the magnetic condition ; the deterioration in this quality of the magnet increasing, as the heat is increased, and becoming very injurious to permanency when the temperature is raised to near the boiling heat of oil.

Column vii., in each of the tables of the series commencing with xxix., shews the progressive deterioration, with the increase of the temperature, in annealing, as proved by the relative effects of the operation of testing. Beyond the heat of 450°, in the case of cast steel, and of 350 in shear steel, it is observable, the influence of the heat becomes very rapidly deteriorating.

20. That the effect of annealing, at all tem-

peratures, on *pairs*, or on *larger combinations*, of *medium* or *thick* plates, is a general disadvantage to the powers of straight-bar magnets of every kind of steel; the disadvantage increasing with the augmentation of the temperature, and the extent of the combination of the plates or bars. On *pairs* of *thin* plates of best cast-steel, the effect, as hereinafter shewn, is different.

A comparison of columns ii. and vi. in Tables xxix. to xxxiii., will be sufficient to verify this proposition. In the semi-medium plates, Tables xxix. and xxxi., a very small improvement seemed to be produced by the annealing at the lowest temperature; but in the medium plates, Table xxx., an opposite effect took place.

21. That the repetition of the process of annealing on the same plates or bars of originally hard steel (the temperature of the oil being the same or not greater) does not appear to produce any alteration in the magnetical capacity, or any additional reduction of the hardness.

Besides the indications of the truth of this proposition, yielded by the series of investigations described in the annexed chapters, a special experiment was made by the annealing of four different descriptions of plates in boiling oil, a part of which had been previously annealed at the temperatures of 300° to 350°, and another part at 550°. No decided difference in the results, either as to capacity for magnetism, or tenaciousness under the action of the test-bar, was observable.

SECT. V.—*Results on the Magnetical Capacities
and Powers of Steel Plates, or Bars, adapted
for* SEA-COMPASSES, *both single and compound,
whether* HARD, *or whether* ANNEALED *at
various Temperatures, with the Effects of
spacing the Plates.*

These Results apply specifically to needles of 6 inches
and 7·5 inches in length, being in the relation to each
other, in proportional masses, of 1 to 2 nearly. In
most of them, the results here given are deducible
from the propositions previously laid down; but they
are here brought together for a more satisfactory determination of the principles affecting the improvement of
the mariner's compass.

22. That for *single-bar needles* constructed on
the principles heretofore generally in use (hardness or temper of the specific degrees H. T. E. S.
being considered), hard cast steel is the most
effective whenever the weight of the 6 inch bar
exceeds 400 to 500′ grains, or the 7·5 inch bar,
about 1000 grains; but in cast-steel needles of
less weight, the directive energy is improved by
tempering or annealing.

This Result is amply maintained, as to 6 inch needles, by
the illustrative remarks on the Results 2 and 3 of chapter
III. The comparison of the powers of two navy needles,
one of them being the best out of nine, with bars of
my own of hard cast steel, as shewn at p. 45, strikingly
verifies the above proposition, where the navy needles
of 510 and 580 grains weight, had a maximum capacity,

in deviating energy, of 13° 30′ and 5° 35′ respectively, and those on my own principle, of 560 and 656 grains, a deviating power of 19° 30′ and 22° 20′.

In plates or bars of 7·5 inch, the limit, as to the weight at which a reduced hardness ceases to have superiority over what we may call *file hardness*, seems to be greater than that due to the proportion of mass. Such a deviation from *apparent* analogy, however, will prove, I apprehend, to be resolvable into another similar deviation from supposed general laws. Heretofore, I believe, it has been considered as a law of magnetics, that the powers of magnets of different but proportional dimensions are proportional to their masses respectively; but this, according to experiments belonging to a future place of this work, will be found *not to be the fact*. It will be proved, I doubt not, that proportional magnets have *not* powers correspondent with their masses.

23. That for *single needles*—of the weights or dimensions ordinarily used for sea-compasses —the best conditions for directive energy among the various kinds of HARD STEEL examined, are found in this order:—*cast* steel from Bradford iron; *shear* steel, S S. or of *hoop*-L iron; *blister* steel, *hoop*-L; *cast* steel, *hoop*-L;—the cast steel from Bradford iron being the most energetic.

Any *apparent* discrepancy betwixt this proposition and the statements previously made, will be explained by a due consideration of the masses employed. When very *weighty* needles are required, best cast steel will be found available for its general superiority.

The results obtained from the trial of the cast steel made out of Bradford iron were so favourable, as to render it probable, that, for the purpose here contemplated, this steel should be selected. Before, however, this conclusion—derived only from a single set of plates— could be safely adopted, further experiments would be necessary, in order to ascertain whether the obtaining of steel from Bradford iron of like magnetical properties could be always relied on.

24. That for *single compass-needles* of ordinary sizes and weight, hard cast-steel from Bradford iron, hard shear-steel (hoop-L), and cast steel (hoop-L), slightly reduced by tempering or annealing, possess apparently the *best qualities*, as combining in a very high degree the properties of power and fixidity.

The diagrams Nos. 5 and 6 represent to the eye the general truth of this proposition; and a comparison of the Tables in Chapter iv., shews a superiority for single needles of *medium* thickness (such as 7·5 inch needles of 900 to 1000 grains in weight) in the cast steel from Bradford iron, and, next to that, of hard shear steel.

As, however, the Bradford steel, whilst exhibiting a certain superiority over all the other kinds tried in *medium* plates, and also in *thin* 6 inch plates, fell short of the hard shear steel in one set of *thin* plates of 7·5 inches in length,—the degree of reliance to be placed on this description of steel cannot be considered as satisfactorily determined : but as such uncertainty does not attach to certain steels from foreign iron, *hard shear-steel*, from hoop-L iron, might with confidence be adopted for compass-needles, or cast steel, of like quality, annealed, as described in proposition No. 27.

25. That plates or bars of hard *shear steel*, suitable for compass-needles, whether of the single or compound form, are injured in their magnetical capacities, if of masses *greater* than those of *thin* plates, by annealing in boiling oil; nor is the very small advantage which may be gained in capacity, in the case of *thin single* plates of shear steel, by annealing at any of the temperatures betwixt 400° and 600°, at all adequate, as a compensation, for the loss in fixidity.

This proposition, which is deducible from Result No. 18, and Table xxxii., is here given, because of its importance in the management of shear steel for compass needles. And the conclusion to which it leads is—that, where shear steel is employed in compasses, the original hardening should not be interfered with, except so far as may necessarily take place in the adjustment of the plates into form.

26. That whilst the extent of the reduction of hardness by the annealing of compass needles in *boiling linseed oil*, is generally injurious to the magnetical properties of *shear-steel* plates, —the same measure of reduction is likewise injurious to the energy of *cast-steel* (hoop L), when the total weight exceeds about 400 grains in 6 inch plates, or about 1000 grains in 7·5 inch plates.

The injurious effect of this extent of reduction of the hardness of *shear steel* is strikingly shewn in diagram

T

No. 6. The powers of a combination of plates of *cast
steel*, tempered, (*CL.* T.), and of the same annealed
(*CL.* A.), when projected in a diagram not here given,
were found extremely similar, almost coincident, in the
case of 6 inch plates of about 140 grains each, even up
to a weight altogether of about 600 grains. With
plates of 7·5 inches long, however, the similarity, though
proximate, was not found to be so exact.

It has been shewn in Proposition 17, that *cast-steel* plates,
of medium mass, gain power by annealing at temperatures *below* the boiling point of oil.

27. That *thin* plates of *cast steel*, adapted for
the needles of compasses, whether employed
singly or in pairs, have their magnetical capacities *improved by annealing* at any of the various
temperatures betwixt 400° and the boiling heat,
—the temperature of boiling yielding the highest
capacity in thin single plates, and that of 500°
to 550° in pairs of plates.

This result, derived from the experiments of Table xxxiii.,
will be seen to be of much practical consequence in
the construction of the needles of sea-compasses. It
determines, and I think satisfactorily, the degree at
which the plates for compasses, when made of cast
steel, should generally be annealed.

28. That for *compound* needles, say of two
thin plates in contact, cast steel annealed,
CL. A., hard shear steel, *SS.* H., or *SL.* H.,
and hard cast steel from Bradford iron, appear
to possess magnetical qualities remarkably
similar in fitness.

This proposition relates to 7·5 inch plates of about 400
grains, or 6 inch plates of about 100 to 200 grains in
weight. Taking the united weight of the 7·5 inch plates
at 800 grains, we find the curves in diagram No. 6,
of cast steel annealed, hard Bradford steel, and hard
shear steel, nearly coincident.

Comparing again the powers of a pair of 7·5 inch plates
for compass needles of different kinds, as shewn in the
Tables, we find—

		Grains.	Deviation.
Shear Steel, *SS.* H. (xxiii.)	total weight	800	21° 25′
Bradford Steel, *Cb.* H. (xxvi.)	„	840	21° 32′
Cast Steel, *CL.* A. (xxvii.)	„	848	21° 33′

In the case of 6 inch needles in pairs, of *less* weight than
200 grains each (say 100 to 150 grains), the steel from
Bradford iron presents some advantage over the other
kinds. And *cast* steel annealed at the heat of boiling
oil, is very analogous to hard *shear* steel up to a combined weight of about 300 grains; beyond that weight,
both hard shear and hard Bradford steel have a
decided pre-eminence.

29. That as the employment of two or more
plates has always an advantage, in magnetical
power, over a like mass in a single bar; so the
separation or *spacing* of such combined plates is
further beneficial to the directive energy of
compound compass needles.

This proposition, which is a general conclusion from the
previous investigations of chapters i. and viii., Part i.,
is here repeated, simply because of its applicability to
the particular subject now under investigation: viz. the
improvement of sea-compasses.

30. That the quantity of gain by *separation* varies in an augmented ratio, directly with the number of plates combined, and inversely with the degree of hardness.

The first part of this proposition is derived from the investigations of chapter ii. The second part is shewn by a comparison of the quantity of gain, by separation, in combinations of tempered plates, [Tables xiii., xiv., and xv.], and some experiments made with hard plates, which have not been described in the body of this work. In the case of *hard* plates, however, the quantity of advantage by spacing was found to be considerably less—and that for the obvious reason, that as there is less loss on the combination of hard plates or bars, than on those that are tempered, so the gain, by attenuating the intensity of the energy in hard magnets, must necessarily be less.

31. That the spacing of thin plates for compass needles of hard shear steel, or slightly annealed cast steel, whether combined in sets of two or four, affords an augmentation of very sensible importance in the directive energy; but in the case of hard cast steel the quantity of gain is not considerable.

In the case of *tempered* plates [*C*. T.] we find an average gain in directive power of about 5 per cent. by the moderate spacing of two very thin plates, and of about 15 per cent. in the spacing of four such plates. Tables xiii., xiv. The tempered plates of 7·5 inches, and about 231 grains weight each, Table xv., were found susceptible, in a combination of four plates, of a greater

power when separated by discs of about one-eighth of
an inch thick, than when in contact, in the ratio of
the tangents of deviation of 25° 22′ to 21° 12′. And
two *annealed* plates, *C. A.*, of 427 grains each, spaced
for a compass needle, received a power of 26° 40′,
which, in contact, was reduced to 22° 46.

But the power of four 6 inch hard cast-steel plates of 123
grains weight each, which in close combination produced, when fully magnetized, a deviation of 19° 45′ on
the compass, was increased only to 20° by separating
the plates into two pairs, about a quarter of an inch
asunder. In like manner four other hard cast-steel
plates of 136 grains, were found to obtain an increase
in power only from 18° 20′ to 18° 50′, scarcely 3 per
cent., by a similar separation. Hard *shear-steel* plates,
however, in accordance with general analogy, gained
considerably more by the separation.

Hence, in the employment of *very hard plates*, in so exceedingly limited a combination, the advantage of total
separation is found to be so inconsiderable as to render
it of little practical consequence whether the plates,
separated in the middle, are united at the extremities,
or kept separate throughout their length.

The further consideration of this subject,
which may justly be held, I conceive, to be of
national importance—involving as it does, the
safety of the lives and property depending on
the guidance of the compass—is reserved for the
practical application of the principles, thus far
determined, in a subsequent part of this work.

The best form, or forms, for the compass needle, will then also fall under consideration; when the disadvantages of the ordinary kind of single-bar needle will be seen, and the variety of forms, with their respective advantages, in which the needles of the compound kind may be conveniently constructed, will be described.

SECT. VI.—*Results in respect to the peculiar effects of hardness and annealing on Magnets of the* HORSE-SHOE *form.*

32. That whilst denomination and quality, degree of hardness and measure of combination or mass, are, severally, qualities of varying influence in their magnetical relations, changing the comparative resultant energy with the changes in these qualities respectively—so the *form* of the magnet (as to straight-bar, or horse-shoe) has likewise its peculiar effect in again influencing the results of hardness, and measure of combination.

> This fact is the general conclusion from the investigations of sect. III. of chapter VI.,—a result, however, which could scarcely have been anticipated from any known analogies.

33. That the conditions requisite for highest energy in large masses, or powerful combinations, in magnets of the *straight-bar form*, are,

as respects the influence of extreme hardness, totally different in the case of magnets of the *horse-shoe form.*

The large combination of hard bars of the horse-shoe form—to the extent of fifteen in number (p. 238)—indicated, in its unexpected feebleness when magnetized and put together, the existence of this peculiar law.

34. That the *extreme hardness* which contributes to the maximum power of considerable combinations of, or masses in, magnets of the straight-bar form, is injurious to the power of a like combination of magnets of the horse-shoe form; and that the annealing or tempering which detracts from the united energy of the straight-bar series, improves the power of the horse-shoe series.

The last experiment described in sect. III. chap. VI., is, I think, quite demonstrative of the truth of this proposition: for in that instance the comparison was made by the use of bars in all respects exactly the same, except as to form,—yet the results, as to their magnetical powers, were totally different. The *extent* of combination to which this law applies, has not been determined.

35. That whilst a given measure of hardness throughout improves the sustaining power of compound magnets of the horse-shoe form; the improvement, in the case of cast steel, attains to a maximum considerably below the condition of extreme hardness.

The hardening of the bars of a large horse-shoe magnet

up to a certain degree—about that perhaps of spring temper, except at the curve, which was still left soft—produced, as has been shewn at page 173, a very great improvement in the energy of the entire magnet; but the hardening of a 15 bar magnet, up to the state of files, as has also been shewn, was very disadvantageous to the power of that instrument.

36. That, for cast-steel magnets of the horseshoe form, and in masses of the ordinary dimensions, the effect of annealing, on hard bars, is extremely beneficial, both to the energy of the bars separately and in combination,—the most beneficial temperature, apparently, being for single bars about 550°, and for bars designed for powerful combinations from 480° to 500°.

It has been shewn already, that the effect of annealing at the temperature of 500° on the average sustaining power, separately, of the bars of two horse-shoe magnets of hard cast steel constructed by Messrs. Stubs, was nearly to treble their original capacity; and that, in combinations of 5 bars and 15 bars, the effect of the same annealing at the same heat was to double their respective powers.

37. That *hard shear steel*, not reduced by tempering, is admirably adapted for the formation of magnets of the horse shoe form—yielding, especially in compound magnets, very superior powers.

This result, which from the foregoing facts was fully anticipated, was, so far as a single trial may be relied on, most satisfactorily determined. For this trial I obtained from Messrs. Stubs, a five-bar magnet of best shear steel, made quite hard, of very beautiful workmanship. Each bar is an inch broad, 0·35 inch thick, and 7·75 inches long, from the extremity of the curved part to that constituting the poles. The total weight is about 9 lbs. The polar extremity is finished quite square—all the bars being of the same length. The conductor presents a surface to the magnet of 2.25 inches by 0·6 inch, and weighs 2300 grains. Each bar, when fully magnetised, sustained an average tension of 10·75 lbs. And when the instrument was put together, it was found to support a load of 58 lbs. before the conductor separated. A larger portion of this great power was retained than I had found in any other instance,— for, on repeated trials, made at various periods, a load of from 50 to 53 was regularly sustained.

The great variety of propositions into which the foregoing Investigations have become resolved, will be sufficient to justify a preceding statement, that no universal answer can possibly be given to the question—What is the best kind of steel, and the best degree of hardness, or mode of tempering, for magnetical instruments? At the same time, we may derive out of these numerous results, and the variety of specific cases to which they apply, a useful summary for the guidance of the practical magnetician in the leading objects of common requirement.

Thus, to speak in general terms, we should

recommend for all *large* or *massive* SINGLE MAG-
NETS of the straight-bar form—the best *cast
steel*, made quite hard : for STRAIGHT-BAR COM-
POUND MAGNETS, generally, the same steel and
hardness : for compound magnets of my BUSK-
PLATE description, the best cast steel, hardened
to the utmost in oil : for HORSE-SHOE MAGNETS,
if *single*, cast steel annealed from file hardness,
at a temperature of about 550°, or shear steel
a little reduced ; and for COMPOUND HORSE-SHOE
magnets, cast steel annealed at 480° to 500°, or
shear steel perfectly hard : for COMPASS NEEDLES,
if *single* and *heavy*, (such as are suited for
stormy weather), hard cast steel ; if *light*, or of
moderate weight, whether single or compound,
best cast steel, annealed at 500° or 550°, or
hard shear steel, or hard cast steel from Bradford iron : and for VERY LIGHT NEEDLES or other
small magnets, the best cast steel annealed,
as with advantage it may be, at the heat of
boiling oil.

In all these cases, the steel, whatever be the
denomination, should be the produce of the
best qualities of foreign iron, except in the
instance of cast steel for compass needles of a
light description, where steel from the best Bradford iron might, it appears, be advantageously
employed.

CHAPTER VII.

ON THE POWERS OF HARD STEEL BARS COMBINED IN VARIOUS WAYS, AS ALSO IN PROPORTIONAL MASSES, SO AS TO FORM COMPOUND MAGNETS OF LARGE DIMENSIONS AND OF GREATER LENGTHS THAN THOSE OF THE ORIGINAL ELEMENTARY BARS; WITH A PRELIMINARY INVESTIGATION CONCERNING THE METHODS OF DETERMINING THE RELATIVE POWERS OF MAGNETS OF DIFFERENT LENGTHS.

THE great advance of magnetical science within the present century, and the surprising results of galvanic action, in the development of magnetic energy in common iron, have led to applications of these results, for practical objects, at once beautiful and useful. And as the present state of magnetical science holds out the prospect of still further applications of this wonderful agency, and, *possibly*, the *useful* employment of the magnet as a moving power; the augmentation of the powers of permanent magnets becomes a desideratum of obvious importance.

In magnets of soft iron, whose energies are excited galvanically, or by means of chemical forces—there appears to be no definable limit to their powers. But as for practical and economical purposes, generally, the waste of materials and employment of acids become at once inconvenient and expensive,—there might be an essential advantage if *permanent magnets* could be obtained of very high degrees of energy.

To this consideration my attention was now, in a particular form of inquiry, directed.

In all the foregoing investigations, on the powers of *combinations* of magnetised steel plates or bars, only one general mode of combination had been examined ; viz. that of a pile, or fasciculus of corresponding pieces. This mode of construction, however, on account of the practical difficulties in the hardening of very large masses of steel, or in making very long bars quite hard and at the same time keeping them fair and in the required form—is not capable of an indefinite enlargement. Hence I proceeded to inquire into the practicability of combining, to a great extent, comparatively small and manageable bars.

This principle of construction for compound magnets of great power, has, indeed, been long since tried, and, to some extent, successfully. But no investigation, that I am aware of, has

been made of the actual powers of a mass of
small magnets so combined, compared with
those of a corresponding solid mass; nor of
the proportional powers of magnets of different
lengths, made up of a single line of similar
integrant bars; nor of the proportional power
of several bars, combined in various ways and
in greater lengths, and the same laid together in a single pile, or fasciculus; nor of the
relation of the ultimate powers of compound
magnets of similar masses, form, and dimensions, when constructed of bars of different
lengths; nor of the relative powers of various
combinations of bars of the same description
(except as to the length of the integrant bars)
in proportional masses; nor as to the possible
extent, as to mass, to which such combinations
may be usefully carried.

These several particulars, however, were
required, in the investigation herein contemplated, to be examined, in order to the effectual
application of the results to the construction of
magnets of extraordinary magnitude and power.

In order to compare the relative powers of
various masses of magnetised steel, or fasciculæ
of magnets, however, accurately with each other,
when their lengths should be different,—some
mode of determining their respective powers,

of approved accuracy, would require to be employed.

That the method of deviations, herein generally adopted, is, in itself, at once very convenient in practice, and, if corrected, in the case of considerable angles, for the equations mentioned in Part I. chap. III. accurate,—has, in its application to magnets of *corresponding lengths*, been already shewn. It has also been stated (page 21) that for bars not equal as to length, their deflecting energy, if ascertained at distances proportionate to their lengths, will generally give results of useful approximation.

The *measure* of approximation to the truth of the method of deviations, for the comparing of the powers of magnets of different lengths, was of importance to be ascertained, in order that a method so extremely ready and effective might be employed in the immediate object of inquiry. The principle of the method, indeed, led me to expect that, provided the deviations could be taken at such a distance from the compass that the angular direction of the forces of the several magnets to be compared, acting on the poles of the compass-needle, should not be very greatly different,—the method of deviations would, though uncorrected, be equally satisfactory for the determination of the relative powers of magnets of *different* lengths, as it has been proved to be susceptible for magnets of various powers of the *same* length. And in this expectation, I was not disappointed.

The most unexceptionable way of determining this point seemed to be by comparing the method of *deviations* with that of *torsion*. For this purpose, I constructed a " balance of torsion," (on the principle originally introduced by Mr. Mitchell, and employed with so much success by M. Coulomb,) of a large size.

The apparatus was made strong and firm, so as to be capable of testing the power of bars of two feet in length and of many pounds weight. The wire employed for this investigation, (except in the first case, which was thinner) was thirty-six inches in length, and about one-sixty-fifth of an inch in thickness. It was obtained of an admirably elastic kind, by the adoption of a hard-drawn wire made expressly for springs used in the Jacquard Loom. A lantern-case for protecting the magnets from the action of the air, with sliding glasses at the upper side and at one of the ends, was provided of sufficient dimensions for the reception of the two-feet bars. To the lower extremity of the torsion-wire was attached a spring stirrup of considerable breadth, calculated to embrace the bars to be tested very firmly, and to secure their perfect parallelism with the line of Zero of the instrument. A

simple mode was also adopted for loading the
lighter magnets, so that the quantity of torsion
might be always determined under similar circumstances as to the load on the wire. A variety
of minor contrivances were adopted for facilitating the repetition of experiments, and
securing accuracy in the comparison of magnets
of different sizes and masses.

With this effective and satisfactory apparatus,
I proceeded to the comparison of the methods of
Torsion and of *Deviations*, by employing them
severally in experiments on the relative Directive powers of Bar magnets of different masses,
lengths, and degrees of energy—with the view
of applying the method of deviations thus experimentally adjusted and verified, to the inquiry
specially contemplated in this chapter. The
leading inquiry, however, being conveniently
extended into *three Cases*, each having an important bearing on the satisfactoriness of these magnetical investigations generally,—I have thought
it desirable to describe them all in this place.

CASE I.—*Comparison of the Powers of various Bar-magnets of the same length (viz. six inches), as determined by the methods of Torsion and of Deviations, in order to the further verification of the method of Deviations.*

The applicability of the method of deviations for the accurate determination of the directive powers of different magnets of *equal* lengths, has been shewn in Part I. chapter III. As, however, the soundness, in science, of this very convenient mode of trial, has been questioned,— the verification of the method by another mode which has met with almost universal approval, will be held, I trust, as quite conclusive.

Eight plates and bars, or combinations of plates, were employed in the experiments on this first case, differing in their powers to the extent of 10 to 1. They comprised the following varieties:—

1. A *pair* of compound bars, each bar consisting of twelve hard cast-steel plates, combined together.

2. The same reduced in energy by one thin magnetic plate interposed in the reverse direction, as to its polarity, betwixt the compound bars.

3. The same further reduced, and in a similar

U

way, by the interposition of two thin magnetic plates with reversed poles.

4. *One* of the compound bars of twelve plates.

5. A single bar of hard cast steel,—a square prism of half an inch thick.

6. A single hard bar, half an inch broad and one-eighth thick.

7. A single tempered plate, half an inch broad by one-sixteenth thick.

8. A single thin compass-needle, three-eighths of an inch broad.

Comparison of the method of *Torsion*, and that of *Deviations*, for the determination of the relative Directive Powers of Bar-magnets of different masses and degrees of energy, made with a series of six-inch Magnetic Bars, Plates, and Compass-needles.

No. of Magnet.	Weight in Grains.	Powers at 4 Lengths from Compass, or 24 Inches.					Powers by Torsion.	
		Mean Deviation.	Tangent.	Ratio, No.1=100.	Equated Tangent.	Proximate Ratio.	Degrees of Torsion.	Ratio, No.1=100.
I.	II.	III.	IV.	V.	VI.	VII.	VIII.	IX.
		° ′						
1	4075	17·58	324	100·0	314	100·0	48·0	100·0
2	4075 - 318	14·17	255	78·7	251	80·0	38·4	80·0
3	4075 - 640	13·15	236	72·8	234	74·5	36·2	75·4
4	2040	9·46	172	53·1	171	54·5	26·0	54·1
5	2950	7·22	130	40·1	130	41·4	19·5	40·6
6	658	4·0	70	21·6	70	22·3	10·8	22·5
7	318	2·59	52	16·1	52	16·6	8·0	16·7
8	145	1·50	32	9·9	32	10·2	5·0	10·4

The analogy of the ratios in columns v. and ix., the ratios deduced directly from the experiments, is sufficiently close to prove the general utility of the method of deviations for the determination of the relative directive powers of magnets of various powers,—the mean difference of the two methods on these eight trials, amounting only to about one-fifty-seventh part of the entire power of each magnet. But, as it will be observed that the whole series of powers, as determined by either method, exhibits differences constantly on the same side (the ratio of column v. being constantly lower than that of column ix.), the application of a small equation to the tangents of the deviations, is here found to be necessary.

This equation (taken very nearly in the proportions indicated in the diagram, in plate ii.) being applied to the tangents of the observed deviations, column iv., we obtain the series of equated tangents, column vi. The proximate ratio by the method of deviations, when thus equated, column vii., agrees so well with that obtained by the method of torsion, as to verify, in the most satisfactory manner, that process for the determination of the powers of bar magnets,—a process on which the results in this work, mainly and essentially depend.

Case II.—*The Determination, experimentally, in Magnets of different lengths, of the relation of their respective deviating effects (at distances from the Compass proportional to their several lengths) to their* ACTUAL DIRECTIVE *Powers.*

The object of this inquiry was to ascertain, how the deviations produced by magnets of different lengths might be reduced, so as to shew the comparative directive powers, or the actual powers with reference to any given standard. In order to this, the determination of the actual powers by the method of torsion, of a series of magnets producing equal angles of deviation at distances from the compass proportionate to their respective lengths, constituted the simplest case which could be well devised.

For this experiment, three straight-bar magnets were employed, of the lengths of 6, 7·5, and 15 inches, respectively,—being, to each other, in the proportions of 4, 5, 10.

Not being either of proportional masses, or of equal deviating powers when magnetized fully, the lowest energy in the three (7° 17′ deviation) was adopted as the standard, and the magnetisms of the other so reduced and adjusted that, at distances from the compass corresponding with four times their several lengths, their deviating powers became alike, or so nearly alike, that the greatest difference from the mean power was only about 2′.

These bars were then placed in succession in the balance of torsion, when their several directive powers were represented by 19·5, 38·5, and 292,—these numbers being the degrees through which the index of the balance required to be turned and the wire twisted, for producing an equal deviation, viz., 30°, from the direction of the magnetic meridian.

Now the proportional powers, thus obtained, just as was anticipated, are found to approximate nearly the cubes of the lengths of the series of magnets. For whilst the cubes of the lengths constitute the series 1, 2·0, 16·0,—the several observed powers of the magnets, reduced to the like extent, constitute the proximate series 1·013, 2·0, 15·2. The differences in the first and last numbers are accounted for by the fact, that the supposed fifteen-inch bar was found to be somewhat short of the proper length, and the six-inch bar a little longer.

For the further confirmation of the law, thus indicated, — a selection was made from other experiments on a more convenient series of bars of six, twelve, and twenty-four inches in length. These afforded the more simple ratio of 1, 2, 4.

Their deviating powers, indeed, were not equal, being 17° 5′, 14° 15′, and 13° 45′; but, guided by the conclusions of case I., there was no difficulty in reducing the powers of the torsion balance to a common standard, by the application of the principle of simple proportions to the observed torsions. Taking, then, the deviation of the middle bar, 14° 15′ (tangent 254) as the standard, the observed torsions, viz., 9·4, 59, and 476, become, when reduced, 7·7, 59·0, 493·5, constituting the series 1, 7·7, 64·1—being a very close approximation to 1, 8, 64, the series formed by the cubes of the lengths of the three magnets. The differences, indeed, are not greater than what may reasonably be attributed to the neglect of the equation of deviations and the errors of observation.

Hence, it is clear that the deviating power of magnets of different lengths, affords, *in the case of equal*, or very nearly equal, *deviations*, a very practical and accurate mode of determining their magnetic powers respectively, -- their powers being, in such case, proportional to the cubes of the lengths of the magnets.

But the case under consideration, as stated in the outset, requires further investigation as to magnets of different lengths having *unequal deviations*. Combining the results, however, of case I., with the above of *equal deviations*, (case II.), the rule becomes general under this simple expression:—

$$P = l^3 t$$

P being the actual directive power of the magnet, t the tangent of the deviation produced at a given distance, proportional to the length of the magnet, from the compass, and l the length of the bar.

The verification of the rule by actual experiment will be abundantly yielded by the descriptions of the next case.

CASE III.—*The Comparative Capacities for Magnetism of equivalent, or similarly proportioned, Bars of Steel of different lengths, as determined both by the method of Deviations and that of Torsion.*

This investigation involved other results. In addition to the object specially proposed for inquiry,—the required confirmation of the rule above stated was looked for. For if the ratios of powers of magnets of different lengths and unequal deviations as yielded by the above formula, should correspond with the ratios by the method of torsion—the rule would derive such confirmation as might be fairly considered to be completely satisfactory.

For the primary investigation, I selected out

of my large stock of magnets, five hard cast-
steel bars of different magnitudes, and of the
respective lengths of thirty-six, twenty-four,
twenty-four, fifteen, and twelve inches. Having
ascertained the weights of the several bars, I
then provided a series of six-inch bars of hard
cast steel, of *equivalent masses* and *weights*,—
the weight of each larger bar being proportional to that of its six-inch equivalent in the
ratio of the cubes of their respective lengths.
Thus, the thirty-six-inch bar, being six times
larger in each of its dimensions (in length,
breadth, and thickness), than its six-inch equivalent, would be heavier than such equivalent
in the ratio of $6^3 : 1^3$, or 216 : 1. For the large
bar, therefore, which weighed 44,950 grains,
the equivalent was a six-inch bar of 209 grains:
for that of twenty-four inches of 43,450 grains,
one of 679, etc.

The whole of these bars being now magnetized
to their utmost capabilities by the processes best
adapted for the full development of the magnetic
energy of the bars respectively,—their several
deviating powers, at four times the length of
each bar, were first ascertained, and then their
respective powers in the balance of torsion.
The various results, with all necessary information respecting the several bars, are exhibited at
one view in the annexed Table.

TABLE of EXPERIMENTS for determining the Comparative Capacities for Magnetism of equivalent, or similarly proportioned, Bars of Steel of different Lengths, both by the Method of Deviations and that of Torsion. As the Degrees of Hardness and the Qualities of Steel, however, of the different bars, herein submitted to Comparison, were not always Correspondent, the Results can only be considered as Proximate.

No. of Experiment.	Bars used in the Experiments.		Equivalent of Bar of 6 Inches.		Comparison of Powers by DEVIATIONS.					Comparison by TORSIONS.			
	Length in Inches.	Weight in Grains.	Equivalent Weight.	Proportion.	Maximum Power of Larger Bars.		Maximum Power of equivalent 6-Inch Bars.		Ratio of Bar to its equivalent of Six Inches.	Torsion of larger Bars.	Torsion of Equivalent Bar of 6 Inches.		Ratio of Capacity of Larger Bar to its equivalent of Six Inches.
					Deviation.	Tangent.	Deviation.	Tangent.			Proportional.	By trial of Six-Inch Bar.	
I.	I.	II.	III.	IV.	V.	VI.	VII.	VIII.	IX.	X.	XI.	XII.	XIII.
					° '		° '						
1	36	44950	209	216:1	7·43	136	13·14	235	1:1·73	[875]	—	—	—
2	24	43450	679	64:1	13·45	245	24·26*	454	1:1·85	476	7·44	13·4*	1:1·8
3	24	20080	314	64:1	9·2	159	17·5*	307	1:1·93	303	4·74	9·3*	1:1·96
4	15	12200	781	15·625:1	21·12	388	28·40	547	1:1·41	177	11·4	16·0	1:1·4
5	12	2670	334	8:1	14·15	254	18·47	340	1:1·34	59	7·4	9·5*	1:1·29

The numbers marked with an asterisk, have been slightly altered, to adjust them, proximately and proportionally, to small differences betwixt the actual mass of the six-inch bar employed for the comparison, and the proportional mass, as shewn in column III., "of equivalent weights." The bar used in experiment No. 3, it should be observed, was of a clearly inferior quality of steel to that of its equivalent,—hence the undue difference in the proportional capacity of the two bars compared.

Various results of much interest and importance are derivable from this set of experiments.

1. Taking the method of Torsion as yielding unquestionately the true proportions of the relative powers of the different bars, we find, by comparing columns XI. and XII., that the power of each of the six-inch bars (column XII.) was much greater than the proportion simply due to its mass (column XI.), as compared with the energy of the large bar of which it was the equivalent. In other words, that the small bars were in every case *stronger proportionally* than the large ones—and the more so, in progressive measure (with the exception above accounted for), as the masses increased. Thus whilst the six-inch equivalent of the twelve-inch bar was a little more than one-fourth stronger, propor-

tionally, than the bar with which it had to be
compared, the six-inch equivalent of the twenty-
four inch bar was above three-fourths stronger
than its proportional one. And these results
are seen to be most satisfactorily verified by the
ratios, column IX., derived from the method of
deviations.

2. Hitherto the degree of accuracy of the
method of deviations, as applied universally to
magnets of every degree of energy, extent of mass,
or variety of length, had been but matter of
inference; and the unimportance, practically,
of the several sources of error induced by the
length of the needle of the compass, which, for
strict accuracy of comparison by direct experiment should be indefinitely short, was matter
of presumption. It was, therefore, important to
ascertain by actual experiment, whether the proportion of error induced by the *varying directions*
of the forces of the magnets under trial with
regard to the poles of the needle of the compass
employed, together with the error for which the
equation already determined provides,—might
be such as to destroy the satisfactoriness of this
method of determining the powers of magnets.

This last set of experiments, however, affords
a satisfactory verification of the practical utility
of the method of deviations, as applied *generally*

to the determination of the relative directive powers of bar magnets. The practical measure of accuracy may be sufficiently inferred from the analogy of the ratio in column ix., determined by the method of deviations, with that of column xiii., derived from the method of torsion. In no case do the discrepancies amount to a twenty-fifth part of the whole power, though in these discrepancies are involved every other source of error in the use of the instruments employed, comprising errors of adjustment and observation, etc.

But a still more satisfactory verification is obtained by comparing the powers of the different magnets employed, as derived from the two methods of deviation and torsion, when reduced to a common standard. For this reduction, I shall take the mean of the powers as given by the six-inch equivalent bars (the last four experiments in the series), viz. 412 tangent, corresponding with 12·05 torsion, and make these amounts, in their respective class, the unit. Applying now the formula given in case ii., we find that the cubes of the lengths, column i., multiplied into the tangents of deviations, column vi., afford, as the result of the method of deviations, the following series,—the mean power of four of the equivalent six-inch bars, or tangent 412, being called 1,—

6 Inch.	12 Inch.	15 Inch.	24 Inch.	24 Inch	36 Inch.
COL. VIII.	No. 5.	No. 4.	No. 3.	No. 2.	No. 1.
1	4·9	14·7	24·7	38·1	71·3

By the method of torsion, reduced in a similar manner (the mean torsion of the equivalent six-inch bars, column XII., or 12·05 being called 1) this series is obtained,—

| 1 | 5·0 | 14·7 | 26·8 | 39·5 | 73·0 |

The series, by deviations, however, is unequated. If, therefore, we apply the equation due to the mean tangent = 12, we have 412—12=400, which being employed instead of 412, affords the following series of still greater correspondence with that given by the method of torsion; viz.—

| 1 | 5·1 | 15·2 | 25·4 | 39·2 | 73·0 |

This close and satisfactory analogy betwixt the ratios of powers, reduced to a common standard, as determined by the two methods of deviation and torsion—not only proves the sufficient accuracy of the former method, for most practical purposes, when applied generally; but this analogy verifies the formula, because of its employment in the reduction of the experiments, by the method of deviations, from which the comparison is derived.

Whilst thus we have so amply verified the practical utility and accuracy of the method of

deviations for the determination of the powers of
bar magnets generally,—we may take occasion,
by the way, of noticing one peculiar advantage
of this method above that of torsion. The method
of torsion, it is observable, gives, of itself, no
absolute value, nor a value whereby the results of
one instrument can be compared with those of
another; but the method of deviations, affords
results indicative of the actual powers of magnets, by reason of its being capable of adjustment to a scale of universal applicability and
comparison.

In order to this, it is only necessary to fix
upon some particular length of bar and some
particular distance from the compass, and then
the tangent of deviation, in all places where the
dip of the needle is similar, will afford a real
standard of comparison, and the particular
tangent decided on, an unit of the scale. In
the general comparison thus yielded, the dip
of the needle, by which the quantity of deviation is necessarily modified, being taken into
consideration, and, if rigid precision were
required, the intensity of terrestrial magnetism,
the comparison would serve for all parts of the
world.

The practical applicability of the method of
deviations for the determination of the relative

powers of straight-bar magnets, being thus satisfactorily established,—we now proceed, in the
confident use of this method, to the subject
proposed for investigation at the head of this
chapter.

For the purpose of pursuing the contemplated
inquiries on a good scale, I procured from
Messrs. Stubs, a quantity of cast-steel bars,
made from the same ingot, of an inch in breadth
and three-sixteenths of an inch in thickness,
extending to the total length of thirty-six feet.
They were constructed into bars of twelve and
eighteen inches each,—eighteen of the former
and twelve of the latter length. The weight of
each set was the same, viz. 11·31 lbs. They were
all ordered to be as hard as files; some few,
however, were found not to be perfectly hard,
though sufficiently so for the complete satisfactoriness of the proposed experiments.

They were magnetized by various processes
and with different powerful compound magnets,
by the way of experiment; and, ultimately,
because in this case found to be the most convenient and effective method, with two fasciculæ
of busk-plate magnets, placed upright and
nearly together in the middle of each bar,
and then drawn asunder and slipped off sideways at the two extremities at the same time,
preserving throughout an upright position.

This operation was twice performed on both
sides of every bar, separately.

To render the series of experiments more complete and generally useful, the powers of these
bars were tried first separately, then in pairs, end
to end, or in double lengths: all the twelve-inch
bars were also tried in treble lengths, and each
set of bars variously combined both with these
of its own set and intermingled with those of
the other. The object in this latter case was to
determine whether the ultimate power of large
combinations would be the same when bars of
different lengths were united, as when all the set
combined were similar,—as such an union of
different lengths would be very desirable, in
order that any large mass so combined might
have more firmness, and be easier bound
together.

The various forms of combination are represented in plate IV.

The following results, as to the powers
exhibited by the various combinations, were experimentally obtained. The several bars, it may
be proper to state, were invariably remagnetized
whenever they had been exposed to a greater
deteriorating influence than that to which they
were about to be subjected in any new arrange-

ment; and, also, that the powers of the several
bars and their various combinations were always
determined by the measure of their deviations
at the distance of *twice* the length of the mass
or masses, under trial, from the centre of the
compass. Thus the twelve-inch bars were
tried, separately, at the distance of two feet
from the compass; the eighteen-inch bars at
three feet distance; two twelve-inch bars in one
length at four feet distance; two eighteen-inch
bars at six feet, etc.

1. The deviating powers of the twelve-inch
bars separately, [see Plate iv., No. 1.] varied
from 17° 55′ to 21° 0′; the mean power being
19° 17′, equal to tangent 350. Length of each
bar, one foot.

2. A pile of ten twelve-inch bars, [No. 2]
gave a power of 48° 50′=tangent 1144. Total
length of steel bars combined, ten feet.—The
powers in progressive combination were as
follow:—1 bar=19° 5′; 2=27° 20′; 4=39° 20′;
6=46° 20′; 8=48° 30′; 10=48° 50′!—In another
series in which eight of the *best* bars were
selected by testing, the powers in combination
were:—1=19° 5′; 2=30° 20′; 4=41° 0′; 6=
47° 40′; 8=51° 34′. In another series, eight of
the weakest bars being selected by testing, the

powers were:—1=20° 10'; 2=27° 37'; 4=
33° 50'; 6=37° 30'; 8=41° 0'.

3. The deviating power of two bars of twelve
inches each, in a line, [No. 3], being the mean
of nine pairs of bars so combined, was 5° 37'=
tangent 98. Length of bars combined, two feet.

4. Four twelve-inch bars (two in length and
two in breadth, No. 4) gave a power of 10° 45'=
tangent 190; but the same bars when tried with
the two rows separated about two inches and
kept in parallelism, the two bars of each row
being in contact, had a power of 11° 28'=
tangent 203. But the average power of six sets
of these bars, so arranged with the two rows
separated, was less than the above, being 10° 52'.
Length of bars combined, four feet.

5. Eight twelve-inch bars, comprising two in
length, two in breadth, and two in thickness,
produced a mean deviation, when in contact as
No. 5, of 17° 9'=tangent 309 ; but when these
bars were presented to the compass, previously,
with each of the four rows a little separated,
their mean action on the compass was a deviation
of 19° 59'=tangent 364.

6. The mean power of three twelve-inch bars
in contact in one line [No. 6] was 2° 25'=tangent
42. When the continuity of the line of magnets
was broken by slipping the middle bar out
sideways, about an inch, keeping each of the

bars at the same distance from the compass as before, the power fell 0° 16', on an average of four sets of bars ; that is, the mean power of the three bars fell to 2° 9'=tangent 38. But when the series was reunited, the original powers were again restored, except in experiments where the series of three bars had been magnetized together, as one bar, when, in that case, the quantity recovered was, on a mean, 0° 8'. Length combined, three feet.

7. Six twelve-inch bars (three in length and two in breadth, all touching) gave a mean deviation of 4° 21'=tangent 76. Length combined, six feet.

8. Twelve twelve-inch bars (two tiers combined as No. 7), gave a power, in contact, of 7° 50'=tangent 138. Length of bars combined, twelve feet.

9. Eighteen twelve-inch bars, in two tiers in contact (comprising three in length, three in breadth, and two in thickness), produced a deviation of 11° 20'=tangent 200. Length combined, eighteen feet.

10. The mean power of the twelve eighteen-inch bars, separately [No. 10] was 8° 32'= tangent 150. Length of each bar, eighteen inches.

11. The total effect of the whole twelve bars of eighteen inches in length, in one pile, in

contact (as No. 11), was a deviation of 42° 0′= tangent 900. Length of the bars eighteen feet. The powers of this set, in progressive combination in the pile, were as follow:—1 bar=9° 0′; 2=15° 0′; 4=23° 30′; 6=30° 6′; 8=34° 50′; 10=38° 54′; 12=42° 0′.

12. The mean power of two eighteen-inch bars in contact, lengthways, was 2° 26′=tangent 42·5. Length combined, three feet.

13. Four bars of eighteen inches (a double row of two each) gave a mean power of 4° 40′= tangent 82. Length combined, six feet.

14. Eight bars of eighteen inches, or two tiers of No. 13, gave a mean deviation of 7° 57′= tangent 140. But the same tried previously, with all the four rows separated about an inch from each other, gave a power of 8° 25′=tangent 148. Length combined, twelve feet.

15. Twelve eighteen inch bars (two tiers, in contact), comprising two in length, three in breadth, and two in thickness, had a power of 11° 8′=tangent 197. Length combined, eighteen feet.

16. A tier of nine bars of twelve inches, (three in length and three in breadth) with a tier upon it of six bars of eighteen inches (two in length and three in breadth) produced a deviation of 11° 8′=tangent 197. Length of bars combined, eighteen feet.

17. A three-tier magnet, with three bars in breadth, comprising nine bars of twelve inches, and twelve of eighteen, gave a deviating power of 15° 22'=tangent 275. Length of bars combined, twenty-seven feet.

18. A four-tier magnet, three bars in breadth, comprising eighteen magnets of twelve inches, and twelve of eighteen inches, placed on each other in alternate series like that of masonry, produced a deviation of 18° 42'=tangent 338. Total length of the bars combined, thirty-six feet; and total weight, 22·6 lbs.

19. The four-tier magnet [No. 18] being arranged in steps, as represented in the figure, each step being an inch, gave a deviation of 18° 22'=tangent 332, when the distance from the compass (six feet) was measured from the extremity of the nearest tier; and 19° 46'=tangent 359, when the mean position of the poles of the four tiers was taken for the measurement of the distance. The steps being reduced to a quarter of an inch each, the deviations, at the two distances taken as above, were 18° 20' and 18° 40'. Length of bars combined, thirty-six feet.

20. The same thirty bars arranged in six tiers, two bars in breadth, all touching, gave a power of 17° 48'=tangent 321 ; and when afterwards opened into four separate rows an inch apart, three tiers in each row, the power increased

to 18° 48′=tangent 340. Length combined, thirty-six feet.

These various effects of contraction on the ultimate magnetic energy of the masses united, are now available for being applied to the inquiries proposed for consideration in this chapter.

I. *As to the Proportional Powers of Magnets of different lengths, but in all other respects, both as to dimensions in breadth and thickness, and in quality and hardness of steel, the same.*

Coulomb, in investigating this question with reference to magnets cut from the same wire or rod, came to the conclusion that the powers of such magnets are proportional, or very nearly proportional, to their lengths. The foregoing experiments enable us to investigate, very accurately, not indeed this precise case, but the analogous one of the relative powers of *combinations* of magnets of the same size and quality in different measures of length.

Let us, for example, compare the power of two twelve-inch bars of steel in a line, with that of one such bar. The data, for the comparison, are afforded us in Nos. 3 and 1,— constituting the mean results of several experiments with eighteen bars of the same kind. The respective powers, as shewn by the formula in case II., are here in this proportion ;—as $1^3 \times$ tangent $350 = 350 : 2^3 \times$ tangent $98 = 784$; or, as $1:2\cdot24$. But the lengths of the two bars are in the exact ratio of 1:2.

If we compare again No. 6, comprising three bars amounting to three feet in length, with No. 1, of one foot,—we have the proportional powers of $1^3 \cdot 350 : 3^3 \cdot 42 = 350 : 1134$, or 1 to $3\cdot24$. But the lengths and masses are in the proportion only of 1 to 3.

We may also compare the power of a length of two eighteen-inch bars [No. 12] with that of one such bar [No. 10]. Here the proportions are, $-1^3 \cdot 150 : 2^3 \cdot \overline{42\cdot5} = 150 : 340$, or $1:2\cdot27$.

To verify the result thus indicated in each of these experiments, viz.—that in a single series of magnetic bars, connected lengthways, of the description here employed, two or three bars so combined, are *stronger* than a single bar, proportionally to their lengths,—I made another special experiment, with a selection from the best of the series of twelve-inch bars. I took four of these bars, and having magnetised them fully with the bundles of busk-plate magnets, I ascertained their directive powers :—1st, singly, at two feet; 2dly, two in a line, touching, at four feet; and 3dly, three in a line at six feet from the compass. Their mean powers at these several distances were 20° 45′, 5° 55′, and 2° 41′,—

the tangents of the deviations being 379, 104,
and 47.   Hence their relative directive powers
give the ratio of 1^{3}·379, 2^{3}·104, and 3^{3}·47; or
reduced as above, of 1, 2·2, 3·345,—their real
proportions, as to length and mass, being 1,
2, and 3.

This result, however, which at first sight
might seem extraordinary, becomes quite intelligible when connected with another set of experiments with the same bars, made at the same
time.   Whilst the two series of two and three
bars were respectively in their places on the
" trial-board," the contact of the bars was
broken by slipping the middle one out sideways,
in the latter case, and the second bar, in the
other case, to the distance of an inch from each
other, but so that the actual distance of each
bar from the compass was the same as before.
Immediately, as in the former experiments (see
experiment No. 6) the quantity of deviation fell
to 5° 33′, with the two bars, and to 2° 32′, with the
three bars; so that the series of tangents, under
this arrangement, became 379, 97, 44.   But
here we have the entire action of the separated
bars, which, apparently, ought to correspond with
their united individual powers, and so constitute
the exact series of 1, 2, 3.   The series reduced
as before, is found to be but slightly different,
viz., 1, 2·05, 3·13.

The discrepancy here, if it had not been accordant with that observed in other experiments, might have been ascribed to error of observation in the deviations. But analogy suggests that the series *should* be in excess. For as the junction of the bars obviously increased, by their inductive powers, the total energy of the mass when thus favourably united—the induction, *without* contact, which must take place betwixt bars separated only by so small a distance as an inch, might reasonably be expected to produce some degree of influence.

II. *As to the Proportional Powers of Magnets combined as in the figures* (plate IV.), *compared with those in a single pile, or fasciculus.*

The results of combination, when the twelve and eighteen-inch magnets were formed progressively into separate piles [Nos. 2 and 11], enable us to form this comparison very satisfactorily.

Let us take, for instance, the case of the series of *eight* bars, arranged as No. 5 (two in length, two in breadth, and two in thickness), and we find a deviating power of 17° 9′=tangent 309. Now eight of these bars combined as No. 2, gave a power of 48° 30′=tangent 1130. The relation of these powers are $1^3 \cdot 1130 : 2^3 \cdot 309 = 1$ to $2 \cdot 2$; that is, the directive power of these eight magnets, when combined in the form of No. 5, is

more than double that of the same series when
combined in a single fasciculus. And comparing again the form No. 5, with the power of
a select series of the best of the eight bars in a
pile, yielding a directive power of 51° 34′=tan-
gent 1260, we have still the preponderance in
favour of the more extended combination, No. 5,
in the ratio of 1·96 to 1.

Compare we again the powers of eight bars
of eighteen inches when combined in the form
No. 14, comprising double bars in length,
breadth, and thickness, with the same bars piled
up, as No. 11, in one fasciculus. The deviation
produced by the longer combination, each bar
touching, was 7° 57′=tangent 140, and that by
the simple pile was 34° 50′=696. The relation
of these powers, reduced by the formula, is
found to be as 1·61 to 1.

The explanation of this fact is, on considerations suggested in the early part of this work,
very intelligible. In the pile, the magnetic
energy is in a more concentrated condition than
in the more extended combination ; hence the
deterioration is much greater in the pile than in
the other form.

The sum of the powers of the eight bars tried
separately, as appears from the original notes of
the experiments, was 1176 tangent. Hence the
difference betwixt this amount and the tangent

of deviation of the eight bars when in a pile, that is $1176-696=480$, represents the quantity of deterioration produced by this form of combination; whilst the difference, in the other case, of 1176 and 140×2^3 or $1120=56$, shews the small comparative loss suffered by the combination in the more extended form.

III. *As to the relation of the Powers of Bars of twelve and of those of eighteen inches when combined so as to constitute masses of the same form, dimensions, and weight.*

The object of this inquiry was to determine whether, in large magnetic combinations, the resulting energy of masses of given dimensions would be the same, or nearly the same, whether the longer or shorter bars of the same kind were employed.

A reference, simply, to the plate, No. IV. or to the foregoing description of the effects of each combination, will answer, sufficiently for our purpose, the inquiry before us. Thus No. 7, composed of six bars of twelve inches, and No. 13 of four bars of eighteen inches, comprise the same dimensions and masses; and their relative powers, tried at the same distance from the compass, were $4^\circ 21'$ and $4^\circ 40'$. Again, No. 8, consisting of twelve bars of twelve inches, agrees in mass and dimensions with No. 14, consisting of eight bars of eighteen inches, and their relative powers were 7° 50′ and 7° 57′. Nos. 9 and 15, likewise correspond in mass and dimensions; and their respective powers were 11° 20′ and 11° 8′. Compared by their tangents, the series of twelve-inch bars in the order above registered, gave the powers of 76, 138, 200 ; and the equivalent series of eighteen-inch bars, of 82, 140, 197. The greatest difference in any two corresponding masses here amounts only to 7·3 per cent. of the total power—a difference by no means exceeding the unavoidable varieties in the powers of such magnets, however carefully constructed.

IV. *As to the Relative Powers of combinations of twelve or eighteen inch Bars, each kind by itself, and those of equal masses and dimensions composed of the two kinds, intermixed.*

This inquiry was designed to determine whether magnetic bars of *dissimilar lengths* (but alike in their other dimensions and qualities) might not be combined together for the formation of magnets of great masses, with equal advantage as the combinations made out of bars of the same length. For if such an arrangement of alternating points of junction, and of alternating series of magnets of different lengths, should yield as much power as like masses composed entirely of one kind, with their points of junction

all corresponding,—then would there be afforded
a great convenience in arrangement, with the
means for yielding solidity, when very large
masses should be united as one magnet.

The cases numbered 9, 15 and 16 constitute
equal masses adapted for this comparison. And
here it will be perceived, that whilst eighteen
bars of twelve inches gave a power of 11° 20′, and
twelve bars of eighteen inches a power of 11° 8′,
a tier of each kind, or nine bars of twelve inches
with six bars of eighteen inches, gave a power
also of 11° 8′. Hence it would appear, that bars
of like description, of dissimilar lengths, may be
combined together without being productive of
any perceptible disadvantage to the resulting
power of the mass.

V. *As to the Relative Powers of various combinations of Bars of the same description (except as
to the length of the integrant bars) in* PROPOR-
TIONAL *masses.*

Here the object of inquiry was—whether various series of magnetic bars, arranged in masses
proportional in all their dimensions, would yield
proportional powers ; that is, whether the series
of bars corresponding with the numbers 1^3, 2^3,
3^3, etc. (arranged according to figures No. 1,
No. 5, No. 17) would afford powers corresponding with the series 1, 8, 27, etc., or, if not, in
what other ratio?

The result obtained in case III. of this chapter,
naturally led to the conclusion, that the powers
of the larger combinations would be *below* their
due proportions; but as this present inquiry
related to proportional masses in *combined* magnets (a condition which is advantageous to the
general result), and the other in *single* or solid
masses—the investigation became necessary.

Comparing now the series of Nos. 1, 5, and 17,
we have the powers, 19° 17′, 17° 9,, 15° 22′,—
shewing a progressive decrease in the proportional energy. The progress of deterioration con-
stitutes this series,—$1^3 \cdot 350$, $2^3 \cdot 309$, $3^3 \cdot 275 =$
1, 7·06, 21·2, instead of 1, 8, 27, the exact
proportional power.

Comparing again the set No. 14, consisting of
eight bars of eighteen inches, with the power of
the single bar, No. 10, we obtain this proportional ratio,—$1^3 \times$ tangent 150 (single bar), and
$2^3 \times$ tangent 140 (eight bars); or reduced as
above, 1, 7·47.

Looking back for comparison to case III., however, we find the measure of deterioration below
the power due to the mass, much less considerable
in magnets formed by combinations of smaller
pieces, than in the instance of proportional solid
bars. Selecting out of the Table of experimental
results, under case III., the powers of the two
analogous cases of bars of twelve and twenty-four

inches, we obtain by calculation the ratio 1:5·4, instead of 1:7·06, and 1:7·47, as above,—which shews that the advantage, in large masses, is in favour of magnets formed out of comparatively small bars. The case of the thirty-six-inch bar is omitted, because it was much too thin for a due comparison.

VI. *As to the extent to which small Magnets, made quite hard, may be advantageously combined, with a view to the formation of larger artificial Magnets than have yet been constructed.*

The largest combination of bars, employed in these experiments, is that figured in No. 18, which had a power, in comparison with that of a twelve inch integrant bar, as tangent 338×3^3: 350×1^3, or as 26:1; but the real proportion of the two masses was as 36:1. Still, however, the measure of deterioration was by no means so considerable as to prevent the effective combination of vastly larger masses.

Taking the experiments already made on combinations in proportional masses, however, we can obtain, by inference, a very reasonable anticipation of the effects of larger combinations. This is accomplished in the simplest manner, by throwing the cases of proportional masses, Nos. 1, 5, and 17 into a geometric curve in a diagram. Proximately, the curve yielded by the

320 MAGNETICAL INVESTIGATIONS.

powers, 1, 7·06, 21·2, for the respective masses
1, 8, 27, combined proportionally, if prolonged
to the next proportional magnitude of 4^3, or 64,
would give a directive power, apparently, of
about thirty-nine times that of a single bar.

Hence whilst we perceive that the deterioration
of power is such as to discourage the idea of constructing permanent artificial magnets, of *limited
lengths*, indefinitely large ; yet such is the nature
of the curve yielded by the powers of a succession
of proportional magnets, that, if the proportions
of the original bar be continually maintained,
and the length, consequently, duly increased,
there would be no limit to the extent of power
obtained. Moreover, it is obvious, from the data
here afforded, that magnets of very great bulk
and of immense powers, may be constructed of
cast-steel bars hardened to the utmost, on this
principle of combination; and that if the combination were made to assume the horse-shoe form,
which is favourable to the construction of magnets of inferior hardness,—a degree of magnetic
energy might easily be obtained, incomparably
surpassing anything yet accomplished.

The plan of combination, indeed, as is well
known, was long ago put into practice by Dr.
Gowin Knight, in the construction of a magnet,
now in possession of the Royal Society, of extraordinary size and mass.

According to the description given of this instrument by Dr. Fothergill, executor to Dr. Knight, in the Philosophical Transactions, vol. LXVI., p. 591, for the year 1776,—it consisted *originally* of two parts, each comprising " two hundred and forty bars disposed in four lengths, every length containing sixty bars, placed in six courses or layers, in contact with one another; and ten in each course placed side by side, in contact also."*

But one of these " Magazines" of the magnetical machine was destroyed or greatly injured by a fire in the house where it was deposited ; but replaced, on the same principle of construction, by Dr. Fothergill.

The whole apparatus is now enclosed in a strong box, measuring forty-two inches by eighteen and seventeen. The magnets present two poles about six inches out of, and beyond, one end of the box, so that the length of the instrument from the extremity of the concealed end to the extremity of the exposed poles, appears to be about forty six inches.

The ends of the projecting poles measure twelve inches by three, each, and are nine inches apart.

* My reference to this paper was taken from the Abridgment of the Philosophical Transactions by Dr. Hutton, etc. Vol. XIV., p. 117.

Dr. Faraday, in the Philosophical Transactions, for 1832, p. 135, has given a short
description of the instrument, stating, " that it is
composed of about four hundred and fifty bar
magnets, each fifteen inches long, one inch wide,
and half an inch thick." And, as to its sustaining power, it is added, that " when a soft iron
cylinder, three quarters of an inch in diameter
and twelve inches long, was put across from one
pole to the other, it required a force of nearly
100 pounds to break the contact." What the
sustaining power, however, is, by its proper
conductor, is not mentioned.

So far as the instrument is visible beyond the
box, I have inspected it, but could not get access,
at the time, to the interior. Compared with the
great mass of magnetic bars employed in its
construction, however, the power was, obviously,
extremely feeble,—though, doubtless, very much
stronger originally. Comprising, as originally
described, a weight of about 1000 lbs. of steel;
or, by estimation on the data given by Dr.
Faraday, betwixt 1200 and 1300 lbs. weight,—
it ought to carry, if properly constructed, the
weight of more than a ton.

But a curious circumstance in the construction
of the instrument is mentioned by Dr. Fothergill,
which, if actually still existing, must conduce
greatly to the injury of the power of the machine.

Dr. Fothergill states, " that it was found diffi
cult, after the final hardening of these bars, to
preserve among them a perfect equality of size ;
therefore the contact of the sides was perfected
by thin iron plates, slipped in between the braces
and the junction of the ends of the bars."
Now this interposition of iron, or even over-
laying of iron plates, must have been most injurious
to the energy of the magnets—as each portion of
iron, applied to any part except the neutral and
transverse axis, would become magnetic in the
converse direction to that of the instrument, and
so neutralize a portion of the original magnetism!
The only other exception to such an injurious
effect would be if the iron were applied after
the manner of an armature: but the description
given by Dr. Fothergill of the interposed iron
plates shews clearly that they were not disposed
of in this way. Whether these iron plates yet
remain, however, I had not the opportunity of
determining.

GENERAL RESULTS.

THE various points investigated in this chapter have been carried so far towards the conclusions here contemplated, that little more than a propositional summary of the several results will now be necessary.

1. That for the determination of the directive powers of straight-bars magnets of the same length, the methods of *Deviation* and *Torsion*, give, to a degree of correspondence sufficiently close for practical purposes generally, the same results.

When the angles of deviation are but small—say under 10°,—the results by the two methods are so nearly correspondent, as scarcely, if at all, to exceed the unavoidable errors of observation [see case I.]; whilst the method of deviations, both for the comparison of experiments by different persons, and for facility of obtaining results in practice, is incomparably the most convenient of the two methods.

When the deviations are large, as observed at two lengths distance of the magnets from the compass, the results, if great precision be requisite, must be equated by the method described in chapter III., Part I.; or else the deviations must be reduced in quantity, by being taken at a greater distance from the compass.

2. That the relation of the powers of magnets of different lengths, is exceedingly well determined by their action on the compass, at distances proportional to their lengths,—the tangents of deviation into the cubes of the length of the bars, respectively, yielding a ratio indicative of the several actual powers.

> In the case of equal deviations being produced by bars of unequal lengths—the ratio of their powers will be simply as the cubes of their lengths.—See cases II. and III.

3. That hard straight-bar magnets, if combined lengthways, or in a straight line, with dissimilar poles touching, yield powers (within the limits and masses experimented on) somewhat greater, as compared with the powers of the single elementary bars, than what is due, proportionally, to their united length.

> The amount of this increase of proportional energy, in the case of a single line of magnets so combined, is shewn, in different instances, in sect. I.

4. That when similar straight-bar magnets are combined in any extended form—that is in any form in which the length of the mass comprises more than the extent of a single bar—the directive power is always greater than when the same mass and series of bars are placed in a single pile.

The less concentration of the magnetic energy in the extended form, than in the single fasciculus, sufficiently
explains, on principles already laid down, the reason
of the greater amount of energy in that mode of
combination.—See sect. II.

5. That sets of magnetic bars of different
lengths—such as those of twelve and eighteen
inches—when combined in their respective kinds
into masses of the same form, weight, and dimensions, appear to yield similar degrees of energy

Within the limits tried in the foregoing experiments—the
differences betwixt the like combinations of the two
classes of magnets, were so inconsiderable as strongly
to support this proposition—the greatest difference in
three cases of comparison not exceeding 7·3 per cent.
of the total energy, and the average difference being
trifling.

6. That sets of magnetic bars of different
lengths, when combined so as to form several
equal tiers in one mass—appear to yield similar
ultimate powers as like combinations in form,
kind, and dimensions, of magnets of one specific
length only.

As far as this comparison could be carried with the
quantity of bars made use of—this result, of much
importance in its application to the construction of
magnets of extraordinary masses, was very satisfactorily
indicated.—See sect. IV.

7. That the relative powers of magnets, whether
single or compound, when different in mass but
proportional in all their dimensions, are not in

the ratio of their respective masses,—the larger masses being less strong, proportionally, than the smaller.

> This proposition, as to *compound* proportional magnets, is determinable from the experiments described in the foregoing chapter, [sect. v.]—being decidedly proved even under circumstances favourable to the improvement of the power, by reason of the combination of comparatively thin bars. Yet in every case of proportional combinations,—such as those of two bars in length, width, and thickness, and of three bars in each of the dimensions of the compound magnet,—the three-bar series was less strong, proportionally, than that of two bars, and the two-bar series than that of the average of the single bars of which the respective compound masses were composed.

> In the case of *solid* magnets of different masses, but tolerably proportionate in their several dimensions—the result was the same, only the differences greater [see case III.]; the methods of torsion and deviation, whilst giving analogous results, shewing a considerable inferiority in proportional power (with respect to mass) in the larger bars. These comparative experiments shewed, that the powers of the equivalent bars of six inches in length, were nearly twice as great, proportionally, as those of bars of twenty-four inches and upwards.

8. That a compound bar, consisting of a series of magnets in a straight line, *loses* power, if the continuity of the series be broken; whilst a compound mass of more than one line of magnets *gains* power by the separation of the lines or tiers of magnets combined.

The first part of this proposition is shewn by the notes on
experiment No. 6; and the second part by those on
Nos. 4, 14, and 20.

The reason of the apparently different effects here produced
by the separation of the magnets is very intelligible;—
the separation of the bars in favourable continuity, as
in the first case, depriving the magnet mainly, of the
gain it had obtained by induction; but the separation
of the parallel rows or lines of bars, where the conti-
guity of similar poles had an injurious tendency, was
necessarily beneficial.

9. That the arranging of the several tiers of a
large compound magnet in the form of steps at
the extremities, does not appear to be of any
material benefit to the directive power of the
mass,—though, it probably may, to the lifting
power of the projecting extremity.

The experiments given in Nos. 18 and 19, only go to the
indication of the result of the first part of this proposition, where the gain appeared to be very inconsiderable.

10. That whilst magnets of large dimensions
are less powerful, with respect to their masses,
than small magnets to which they are exactly
proportional in all their dimensions; and whilst
the increase of the dimensions continually deteri-
orates from the quantity of energy due to the
mass,—yet magnets may be combined in such
proportional dimensions, with a constant increase
of power, ad infinitum.

This proposition results from the nature of the curve

derived from the projection of the powers of proportional magnets—which curve being of a hyperbolic
character, must continue to separate to a greater distance from the line of abscissæ, however extensively
prolonged.

A very different result takes places, as to the augmentation
for practical utility, when many bars are combined in a
single pile, or arranged within the same original extent
as to length. In this case, as we have frequently seen
in the foregoing Investigations, a maximum power,
as to practical utility, is, whenever magnetic bars are
so combined, soon attainable. And although, by the
employment of hard steel plates, the attainment of this
maximum is greatly protracted, yet in that kind of
magnet, also, there is an obviously increasing tendency
to render the accumulation of plates, after a certain
amount, but little advantageous.

CHAPTER VIII.

OF THE MAGNETICAL POWERS, RECEPTIVE AND PERMANENT,
OF *CAST IRON*, BOTH IN SEPARATE BARS OR PLATES,
AND IN VARIOUS COMBINATIONS OF PLATES.

———

THIS portion of my Magnetical Investigations is here introduced, in order to place the magnetical powers of *Cast Iron*—a substance of a somewhat intermediate nature betwixt iron and steel—in a distinct position of comparison with reference to those of steel.

So long ago as the year 1832, I commenced a series of experiments on the magnetical recipient properties of *Cast Iron* of various qualities, with specimens furnished me by Edward Roscoe, Esq., of Liverpool. These, however, owing to the thickness of the masses employed (being an inch square) were not fitted for decisive results, in regard to the capabilities of cast iron for the magnetic condition. With the view, therefore, of a satisfactory determination of the comparative capacities of cast iron and steel, I recently added another class of experiments with plates of cast iron constructed of somewhat corresponding dimensions with those of the 7·5 inch steel plates chiefly used in the Investigations described in chapter IV.

As the investigation was one of considerable interest, with reference to the applicability of a material so unexpensive, for the construction of permanent magnets,—I obtained specimens of cast iron in considerable variety, and subjected each kind to the usual tests for the determination of its magnetical properties.

These specimens, with the exception of the series of " Run-steel," from Mr. Lucas's works near Sheffield, were obtained at the Bowling or at the Low-Moor Iron Works, in the parish of Bradford, under the kind assistance of the managers of the respective companies.

It may be convenient, in giving the results of this inquiry, first to describe the several kinds of cast iron employed in the experiments, and then to furnish, in consecutive tables, their respective powers.

1. The first set I have to describe, consisted of six plates of Bowling iron, cast out of grey, or No. 1, pig-iron,—usually considered of the best quality, and of a good description of that kind of metal. It is much less hard, being easily scratched with a file, melts easier, and runs more fluid, than the other kinds of pig-

iron. It has a comparatively dark-coloured, coarse-grained fracture. It is smelted from the best ores of the district, which ores are combined with a larger proportion of coke and lime, than in the manufacture of the pig-metal No. 2, or No. 3. The proportions of the materials combined in the furnace of course vary with the relative qualities of the lime-stone and of the iron-stone; but the proportions used in the Bowling Iron Works, in the manufacture of No. 1 pig-iron (taking the average of the seasons of summer and winter), may be considered to be about 7 cwt. of coke and 3 cwt. of lime-stone, to 6 or 7 cwt. of ore. Such is the effect of temperature or of humidity on the process of smelting, however, that the proportions of lime and coke employed in the production of No. 3 iron, in the summer, or in damp mild weather, will generally be adequate for the production of the highest quality, No. 1, in winter.

The six plates of No. 1, cast iron, employed in these experiments, were 7·5 inches long, 0·75 inch broad, 0·133 inch thick, and weighed, on an average, 1230 grains.

2. Six plates of white, or No. 3, pig-iron. This is a very hard description of metal, brittle, and of a whitish silvery-grey fracture. None of the plates of this set were scratched with the file. The proportions of lime and coke usually

employed at the Bowling Iron Works, in the production of this quality of metal, taking the average of summer and winter, are 7 cwt. of coke, and rather less than half the quantity of lime-stone, to 7 or 8 cwt. of iron-stone.

The difference of price in the market betwixt cast iron No. 1, and No. 3, is usually about ten shillings per ton.

The dimensions of the plates of this set were nearly the same as those of the preceding one, except as to thickness,—these plates being 0·14 inch thick, and of the average weight of 1400 grains.

3. Eight plates of metal, No. 1, from the Low-Moor Works. These were rather shorter than the others, being 7·2 inches in length, and 1188 grains, average weight. These were cast hard, from the "Blow-hole," a furnace heated by air thrown into it by an engine. The iron, it is considered, would have been stronger if it had been melted in an air furnace.

The average quantity of materials used at the Low-Moor Iron Works, for making a ton of pig iron, of the average quality and at the mean of different seasons during last year, is stated to have been 87 cwt. of coal, 25 cwt. of lime-stone, and 76 cwt. of iron-stone. Taken at an average, also, as to season and weather, about 15 cwt. more coal is required to make No. 1, than No. 3, pig iron.

4. Eight plates of "gun-metal," from the Low-Moor Iron Works.—This metal consists of a mixture of different qualities or castings of pig iron, selected of a good description of each kind; the object of the mixture being to obtain an union of toughness and strength. The proportions of this mixture vary according to the size and class of guns—lower numbers of metal being required, for producing the requisite hardness, in very heavy guns. The specimens furnished me were from metal prepared by an extra melting for guns for the navy.

These plates, like the former, were of the length of 7·2 inches; the average weight 1220 grains.

5. Six plates of Bowling iron from picked ore, constituting quality No. 1, pig iron, designed for ultimately being made into spring steel. This, could the result of the furnace be always calculated upon, should be a first-rate quality, being the most expensive in production of any iron made of Bowling ores.

The dimensions of the plates corresponded, except as to thickness, with the other sets of Bowling iron. Average weight 1117 grains.

6. Six plates of "*finery iron*," from the Bowling Works, of the average weight of 1418 grains. —This description of metal is not used for casting, in any manufacture at Bowling; the process of

the refinery being employed simply as preparatory for the "puddling furnace," in order to the production of malleable iron.

Whilst the operation of the refinery improves the purity of the metal, it separates a considerable portion of the carbon and oxygen from the general mass of the iron, and yields a characteristic quality which renders the *puddling* process less wasteful and laborious, and more equable and certain in its results.

The metal from the refinery is extremely hard and brittle, close-grained, and almost of a silvery whiteness.

7. Nine plates of "*run-steel*," from the works of Edward Lucas, Esq., of Dronfield, near Sheffield.—These were of the ordinary dimensions, and weighed, on an average, 1500 grains. They were requested to be cast so as to give the greatest possible hardness.

This metal, in its original state of pig-iron, is obtained from the red hæmitite of Cumberland, smelted with charcoal, and so called "charcoal pig." The characteristic quality of run-steel is produced by a patent process for rendering the articles cast from it malleable. This is effected by cementation with the peroxide of iron, by which the superabundant carbon combines with the oxide and "metallizes" it, without undergoing fusion as in the puddling process. Thus treated, the metal is freed from the combination of elements conducing to the more characteristic peculiarities of cast iron, so as to approximate the purity of malleable iron; at the same time it retains so much of its original quality as to be susceptible of a considerable degree of hardness and polish. Hence it becomes available for the construction of a variety of articles, as a substitute for both malleable iron and steel.

8. Six plates of best *cast steel*, quite soft, were employed in this investigation, for comparison with the powers of cast-iron plates. These were of similar dimensions as the foregoing,—length 7·5 inches, breadth ·75 inch and· thickness 137 inch, and the average weight 1400 grains.

9. Three plates of *gun-metal*, Bowling. This iron consisted of a mixture of Nos. 1, 2, and 3 pig-iron, with about a third-part of old cannon melted down. It was only adapted for small guns. Length of the bars 7·5 inches, average weight 1213 grains.

10. Three plates of an inferior iron, belonging to the general denomination, No. 3, pig-iron, being the produce of "white rake-stone," a very inferior quality of ore, and a quality yielding so small a proportion of iron as not now to be worth smelting. The length of these plates was 7·5 inches, and the average weight 1138 grains.

The foregoing specimens of cast iron were, for the most part, taken from the pig or the refinery, and cast in what is technically called "greensand," that is, moist sand, in contradistinction to dry sand, which, unless for special purposes, is ordinarily used for castings. The effect of the moisture in the sand, especially on thin plates, is to render the casting *very hard*, a condition which I was desirous of producing.

After being cast, the plates were ground to a smooth and even surface; and, being numbered with ink marks, were successively magnetized by the usual process, and with the most powerful pair of my magnets. Their powers, both separately and in combination, were then ascertained at two lengths distance from the compass, on the revolving ruler of the trial board, in the usual manner.

In addition to the varieties of cast iron now described, other specimens were obtained, and their magnetical qualities determined; but the particulars concerning them have not been included, because not contributing any thing satisfactory to the object of this investigation. Among these unrecorded experiments was the trial of a set of plates cast from *Lancashire charcoal pig*, kindly furnished me by Messrs. E. Lucas and Son; the magnetical powers of

z

these however, though higher, were not very greatly different from the powers of the run-steel, No. 7. This set was in the rough, just as the plates were cast; but the magnetical powers did not appear to be materially, if at all, affected by this circumstance.

Besides the magnetical qualities exhibited by the experiments on the several specimens of *hard cast iron*, I tried the effect of reducing the hardness by the process of annealing. For this purpose the plates, sets 1 and 2, were placed in an iron box along with a quantity of iron scales from the smithy. The box was then placed in a small furnace and heated to redness, and then closed up and left to cool very slowly during many hours. The plates came out singularly soft, and those of No. 1, considerably ductile. But the magnetic energy, both in capacity, singly, as well as in combination, and also in fixidity, was found to be greatly injured —just in accordance with analogy in the case of the softening of steel.

The following series of Tables will give, at one view, the particulars, the most important to be determined, concerning the magnetical powers of these various sets of plates.

Magnetical Powers of various descriptions of Cast Iron, in Plates of 7·5 and 7·2 inches in length, and 75 inch in breadth; weighing from 1000 to 1500 grains each.

No of the Plates.	Magnetical Powers.		No. of the Plates.	Magnetical Powers.	
	Separately.	In Combination.		Separately.	In Combination.

	I. Cast Iron, No. 1. 7·5 inch.—Weight, 1230 Grains.			II. Cast Iron, No. 3. 7·5 inch.—Weight, 1400 Grains.	
	o ′	o ′		o ′	o ′
1	13·53	13·53	1	10·0	10·0
2	13·52	18·10	2	10·15	12·30
3	13·35	20·34	3	10·0	14·20
4	13·37	22·15	4	10·15	15·40
5	13·46	24·26	5	9·50	16·35
6	13·20	25·47	6	10·23	17·44
Mean.	13·41	——	Mean.	10·7	——

	III. Cast Iron, No. 1. 7·2 inch.—Weight, 1188 Grains.			IV. Gun Metal. 7 2 inch.—Weight, 1220 Grains.	
	o ′	o ′		o ′	o ′
1	13·14	13·14	1	13·20	13·20
2	13·52	18·7	2	13·15	16·47
3	13·30	20·45	3	13·4	18·40
4	13·27	22·30	4	12·50	20·47
5	13·20	24·10	5	12·45	22·0
6	13·14	25·40	6	12·35	23·12
7	13·32	27·10	7	13·10	24·52
8	13·40	28·25	8	13·0	26·0
Mean.	13·29	——	Mean.	13·0	——

No. of Plates.	Magnetical Powers.		No. of Plates.	Magnetical Powers.	
	Separately.	In Combination.		Separately.	In Combination.
	V. *Steel Iron.* Weight, 1117 Grains.			VI. *Refinery Iron.* Weight, 1418 Grains.	
1	10·32	10·32	1	12·29	12·29
2	11·33	15·8	2	12·40	16·16
3	12·18	18·9	3	12·7	18·14
4	11·14	19·24	4	11·45	20·4
5	8·42	20·10	5	12·19	21·36
6	11·53	22·0	6	12·1	23·15
Mean.	11·2	——	Mean.	12·13	——
	VII. *Lucas's Run Steel.* Weight, 1500 Grains.			VIII. *Soft Cast Steel.* Weight, 1400 Grains.	
1	9·40	9·40	1	13·15	13·15
2	10·32	13·10	2	13·20	16·4
3	10·15	14·10	3	13·25	17·37
4	10·10	15·55	4	13·23	18·45
5	10·12	17·0	5	13·12	19·55
6	10·32	18·30	6	13·47	21.50
Mean.	10·13	——	Mean.	13·24	——
	IX. *Gun Metal.* 7·5 inch. Weight, 1213 Grains.			X. *White Rake, No. 3 Iron.* Weight, 1138 Grains.	
1	9·35	9·35	1	8·42	8·42
2	9·15	11·42	2	8·35	11·15
3	9·8	13·14	3	8·58	12·28
Mean.	9·19	——	Mean.	9·19	——

It may be proper to notice, in connexion with
the specimens used in these experiments, that
the characteristic difference betwixt the various
qualities and kinds of cast iron produced in the
works from whence they were supplied—is considered mainly to arise from differences in the
quality of the ore, and from differences in the
proportions of the materials used in the smelting.
Thus, from the greater quantity of coke employed in the production of " No. 1, pig metal,"
there is supposed to result a larger proportion of
carbon in combination with the metal.   Hence,
No. 1, pig metal, has been described as a super-
carbonated crude iron; and No. 3, as a "carbo-
oxygenated crude iron."

These distinctions, however, are not universally
acknowledged; for some chemists consider that
the characteristic differences observable in the
several kinds of cast iron are rather owing to
the mode of combination of the constituent
elements, than in the quantity of carbon.   But
the result of these experiments would have led
me to infer that there is a predominance of
carbon in the quality No. 1, pig-iron.

Whatever be the characteristic quality of iron,
ordinarily produced from any specific kind of
ore, or proportion of materials; this, as we have
already remarked, is subject to considerable
differences, arising from other causes.   Thus,

besides the effects of management in the process of smelting, both the *quality* and *quantity*
of the produce of the furnaces are found to be
essentially influenced by the temperature and
by the hygrometrical and electrical (?) state of
the atmosphere—a low temperature and dry air
being decidedly favourable, and warmth and
dampness, or an electric condition of the air,
being considered unfavourable to the manufac
ture of cast iron.

Reasoning on the analogies with regard to
the magnetical properties of steel, compared
with those of iron,—I had supposed that the
magnetical qualities in cast iron should increase
with the predominance of the carbon, and
diminish both with the diminution of carbon or
proportional predominance of oxygen :—considering that it was probably the predominance of
either of these alterative elements to which, in
considerable measure, the different magnetical
capabilities of the metal were to be ascribed.

This idea, if not fully substantiated, is not
altogether disproved by the foregoing experiments, in which, it is observable, that the
quality of No. 1, pig metal, usually considered
as the most highly carbonized, exhibits, among
the various kinds of iron examined, the highest
magnetical qualities ; whilst the refinery iron,

which clearly possesses less carbon than the others, has lower magnetic qualities.

The proportional energy and fixidity of magnetism in cast iron, it will be seen, are *low* in comparison of these properties in *hard steel;* but not so if compared with plates of like magnitude of soft, or slightly tempered, steel. And it is by no means improbable, but peculiar specimens of cast iron from other ores might afford still higher capabilities for the magnetic condition.

In diagram, No. 5, the magnetical powers of the best cast iron are represented by one of the lowest curves, where they can be compared, by the eye, with the powers of both *soft* and *hard* steel plates of various denominations. The relative powers of plates of *soft* cast steel appear in the lowest curve of the whole, and those of a series of solid bars of blister steel, tempered at the ends, weighing 1580, 2630, and 5280 grains, respectively, constitute the intermediate curve betwixt hard cast-iron and soft cast-steel. The conclusions arrived at from this examination, with a more definite view of the proportional powers and magnetic capabilities of cast iron, are given in the following summary of Propositional Results.

RESULTS.

NOTWITHSTANDING the variety of specimens of *Cast Iron*, subjected to experiment for the trial of their respective magnetic properties, only a very few general results are suggested by these Investigations. The following, however, seem to be fairly borne out by the experiments.

1. That *Cast Iron* possesses considerable powers for magnetism, both in capacity and retentiveness, though greatly inferior in both qualities to those of properly hardened steel.

> A very cursory inspection of the preceding tables will be sufficient to justify this general proposition.

2. That the best qualities of cast iron, distinguished by their higher value in commerce— seem to possess, unless very deficient in carbon, the highest magnetical properties.

> Thus the quality No. 1, pig, which usually bears a price in the market of about 10s. a ton higher than that of No. 3, is seen to possess a great superiority in its magnetical qualities,—see set No. i. and No. ii. So also, Low-

Moor gun metal, steel-iron, and finery iron, exhibit good magnetical qualities; whilst the commoner iron, sets II. and X., and the mixed, set IX., give lower degrees of energy.

The low energy of set IX., was probably occasioned by some accidental cause—either by the casting being less hard, or by being incorporated with a decarbonized quality of old metal.

3. That in any particular quality of cast iron, the hardness produced by the rapid cooling of the casting, is favourable to the magnetic energy.

The experiment on the annealing of sets No. I. and II. referred to at page 338, abundantly proved this reasonably anticipated result.

4. That the existence of carbon in the cast iron, appears to be beneficial to its magnetical properties, and the processes for decarbonizing, therefore, injurious.

The low powers of the "Run steel," set VII., which, in its general qualities promised to be superior, indicates, I think, that this proportion is accordant with fact. And even the case of the refined iron, set VI., the powers of which are very considerable, does not militate against this conclusion; since the refining process, whilst it improves the purity and increases the hardness of the iron, leaves it ultimately of lower magnetical capabilities than before its being thus partially decarbonized.

The experiment with iron, cast from Lancashire charcoal pig (page 337), did not fulfil the expectation formed concerning that quality. It is difficult, however, from any single set of castings, to draw general conclusions; as an accidental peculiarity in the portion of original metal selected, or in the method of casting, might have a considerable influence on the resulting character of the plates.

5. That in relation to steel, the magnetical properties of the best kinds of cast iron, are equal, if not greater than, those of *soft* cast steel [*C.* S.] in single, medium, or thick plates ; and considerably superior to soft steel in large masses or in combinations of several plates.

> Comparing tables I., III., and IV., with that representing the power of soft cast steel, VIII., this result is clearly established; whilst the lower curves in diagram, No. 5, exhibit the fact plainly to the eye.

6. That cast iron of the best magnetical qualities, such as No. 1, pig-metal, if cast very hard in thin bars or plates, appears to be capable, by combination, of forming as powerful magnets as, or even more powerful than, *solid massive bars* of ordinary steel if only hardened slightly at the ends, or in the manner heretofore commonly adopted in the construction of magnets in this country.

> This is a result drawn from a comparison of the powers of the cast-iron plates given in the foregoing tables, especially of sets I., II., and IV., with the powers of a series of solid bars, tempered at ends, E, as shewn in page 343. A result, very nearly accordant, was derived from a like comparison of the cast-iron 7·5 inch plates, with the set of cast steel six-inch bars, E, tempered at

the ends, represented in the diagram, plate ii. The relative powers being reduced to a common standard in relation to the different lengths, it appeared that, in masses below 6000 grains weight, the solid *steel* bars (in the case of magnets of 7·5 inches in length) had the advantage; but the curve of the powers of cast-iron plates crossed that of solid bars, tempered E, at the weight of about 6000 grains, and beyond that weight the cast iron appeared to gain the pre-eminence.

CHAPTER IX.

ON THE MEASURE OF PERMANENCY OF THE ENERGY IN STRAIGHT-BAR MAGNETS, BOTH SINGLE AND COMPOUND, AND OF DIFFERENT DEGREES OF HARDNESS, AS SELF-SUSTAINED, OR AS INFLUENCED ONLY BY TERRESTRIAL MAGNETISM.

———

ALL artificial magnets, especially those in a highly energetic condition, are subject to deterioration of power, unless particularly protected by conductors betwixt contrary poles, or other sustaining influence.

Questions of considerable interest and importance, as concern the use and efficiency of magnetical instruments, naturally arise out of the knowledge of this tendency to deterioration. What may be the rate of deterioration in the energy of magnets, as accelerated or retarded by terrestrial magnetism? what may be the relative effects of various degrees of hardness on the retentiveness of the magnetic condition? and, whether the permanency of *combinations* of magnetic plates or bars be as great as, or greater than, that of single or *solid* bar magnets? The permanency here referred to, it should

be observed, is not as to fixidity under the
action of magnet upon magnet; but that of the
simple sustaining power of the substance of
which the magnet is composed, under the mere
operation of the ordinary tendency in the magnet to return to a neutral condition, or as affected
only by terrestrial magnetism.

The inquiries herein suggested, evidently
belong to the results of continuous action in
the influences tending to deteriorate the magnetic energy; inquiries, therefore, only to be
answered satisfactorily by repeated experiments
made on the same magnets, cautiously laid by,
after considerable intervals of time.

For this purpose, trial of the permanency of
the magnetic condition was made on a variety
of plates and bars, both single and compound;
the plates in some of the latter being separated
throughout by discs of wood, some separated
partially, and others placed together in contact.
And although with the more extensive combinations the investigations were but imperfectly
carried out; yet enough has, I conceive, been
accomplished for shewing the relative conser-
vative power both of the compound arrangement
of magnetized plates (having reference to forms
of combination suitable for the directing magnets or needles of sea-compasses) and of single
bar magnets, during considerable periods of
time.

SECT. I.—In the first instance, I instituted a comparison, in a series of twelve experiments, betwixt different sets of thin two-feet magnetical plates, tempered throughout, six in each set, and separated by discs of about the tenth of an inch in thickness; and four-bar magnets of two feet in length, of considerable power, two of them being tempered throughout, and two only tempered at the ends. The experiments with the series of thin plates, were made at one length distance from the compass,—those with the bars, at two lengths; but the comparison as to the proportion of loss is equally satisfactory. During the interval betwixt the two trials of their powers, the several sets of plates, or bars, were left without conductors, at a considerable distance from other magnets, and generally laid in a horizontal position, *east* and *west*, so as to derive neither support nor injury from the earth's magnetism. And each bar or set of plates, was, for the most part, first tried immediately after being magnetized, or combined in series, so as to be in its most energetic condition.

The annexed table exhibits the results of the experiments for determining the change, or subsidence, in energy of combinations of six magnetized steel plates, of two feet in length, when left, for a specific period, unprotected. The weight of each set of six plates was about 2·5 lbs.

Series of Plates.	Interval of time in days.	Power when first tried.		Power when re-examined.	
		Mean Deviation.	Tangent.	Mean Deviation.	Tangent.
		o '		o '	
A	3	56·15	1497	55·45	1469
B	3	55·25	1450	54·48	1418
A	4	54·40	1411	54·20	1393
A	7	54·40	1411	54·17	1391
A	11	54·18	1392	54·23	1396
B	3½	55·25	1450	54·45	1415
C	3½	54·3	1379	53·5	1331
E	3½	52·32	1305	51·17	1247
A	23	56·3	1485	55·51	1474
B	23	53·57	1374	53·9	1334
F	23	48·50	1144	48·46	1141
B	6	54·43	1413	54·25	1398
Mean	9·46		1392·6		1367·2

Average loss per cent. in 9·46 days,=1·82.

The retentiveness of these various sets of combined plates, as compared with that of single-bar magnets of corresponding length and breadth, was tried by examining the magnetic powers, respectively, of various single bars, after certain intervals of time—commencing with the highest magnetic energy of which they were susceptible.

The experiments for determining the subsidence of energy in four separate two-feet bar magnets of considerable power, when left, for a short specified period unprotected, are herein exhibited. Each of the bars X were of the weight of about 6¼ lbs.; those marked Y were a little under 3 lbs. weight each.

Description of Bars.	Interval of time. Days.	Powers when first Magnetised.		Powers when re-examined.	
		Deviation.	Tangent.	Deviation.	Tangent.
X 1.	3	14·35	260	14·24	257
X 2.	3	15·7	270	14·44	263
Y 1.	3	7·41	135	7·31	132
Y 2.	3	7·45	136	7·41	135
Y 1.	10	7.41	135	7·29	131
Y 2.	10	7·45	136	7·38	134
Sum.	32		1072		1052
Mean.	5·3	Loss per cent. in 5·3 days=1·87			

As far as these experiments go, we find, that the subsidence, or loss of power, in the single-bar magnets wherein the tension (or total energy) was inferior, was even greater in 5·3 days, than that of the combination of plates, in 9·46 days—indicating a permanency, or tenacity of power, in favour, on the whole, of the compound series.

The loss of power per cent. in these experiments, during so short period, is not to be considered as by any means implying that a corresponding loss of power would take place in subsequent similar periods. For in some single experiments, made with other sets of plates, a very remarkable tenacity was observed during far longer intervals of time. The immediate loss of power here observed

is comparatively great, because of the bars or
plates being magnetized beyond their strength for
retention; but if the first tendency to subside be
somewhat aided by the juxta-position of other
adverse or opposing magnetic influence, the power
will at once sink to a certain extent, and then
the subsequent deterioration, from mere subsidence of power, is found to be less considerable.
And when that adverse magnetic influence has
been considerable—such as that of a moment's
contact of the extremity of a bar or series of
plates with the like pole of another magnet—
then the power is found in certain cases to sink
even below the real strength of the magnet for
retention, so as to be recovered in some measure
by being laid in a favourable position, or within
the influence of the converse polarity of another
magnet.   In one example, the fifth in the series
of the first of the tables just given, an increase
of the original power was indicated; but whether
this arose from some such accidental influence,
or from error of observation, I have no means of
satisfactorily determining. But, notwithstanding
the discrepancy of that particular example, I
could not feel justified in excluding it from the
series.

   In order to shew the measure of permanency
in the compound series for more considerable
intervals of time, and likewise the diminish-

ing deterioration after the first, I would next adduce an experiment with two sets of smaller plates, to which the table at page 137 refers. After the experiments specified in that table, each series of plates, twelve in number, was tied tight with twine, and laid aside, being kept carefully clear of all other magnets, as well as of each other. The powers of both series, at two lengths from the compass, previous to being put aside, with their powers on examination after intervals of one day, and of sixty-four days, were as follows:—

	Original Power.		Power after 1 Day.		Power after 64 Days.	
	Mean Deviation	Tangent.	Deviation.	Tangent.	Deviation.	Tangent.
Harder Series, reduced only in the middle.	24·57	465	24·5	447	22·50	421
Softer Series, reduced throughout.	23·21	432	22·32	415	20·30	374

Now these observations indicate a loss of power in the harder, of 3·88 per cent. in the first day, and of 5·82 per cent. in the *next* sixty-four days; and a loss of 3·93 per cent. in the *softer* series, the first day, and 9·88 per cent. in the next sixty-four days. The greater degree of permanency of the harder plates is here very decidedly shewn—which, probably, would have been still more striking had a comparison been instituted with other plates in which the difference of temper was greater. The position of these bundles of plates, when laid up in the intervals of the trials of their powers, was in the line of the magnetic east and west, nearly; hence, in a condition in which terrestrial magnetism became inoperative.

SECT. II. The measure of permanency of the magnetic condition in smaller bars of various degrees of hardness, both single and compound, but chiefly single, was tried for much longer intervals of time.

In one series of experiments, plates and bars, both single and compound, of the length of 7·5 inches were employed. These consisted of a moderately hard bar of shear steel (about the temper T) of 5310 grains weight; three plates of shear steel, weighing about 650 grains each, of different degrees of hardness; and a compound bar of three plates of hard cast steel tied together, weighing nearly 2000 grains.

These five magnets (all being magnetized in the highest degree) were laid in a cabinet, sufficiently separated to guard against any sensible influence from each other, with their marked ends toward the north—or in the direction they would have assumed if suspended on a pivot as compass needles. After remaining about two months, their powers were examined; when there appeared a *very slight* subsidence (scarcely amounting to two per cent. on the whole) of the original energy, being the greatest in the heaviest bar of shear steel. The magnets being replaced in the cabinet, as before, were allowed to remain, with only one trial intermediately, for rather more than a year, when they were again placed in succession on the ruler of the *Trial-board*, and their directive powers determined. In four out of five (the other having been removed during the interval) the powers were so nearly the same as they were a twelve-month before, that the alterations did not exceed the small discrepancies which might be expected to arise from the errors of observation and of the adjustment of the apparatus : for the average deterioration did not exceed a four-hundredth part of the whole power.

A second series of experiments was coincidently proceeded with, in which six-inch plates and bars were employed. These consisted chiefly of selections, of different degrees of hardness, out of the series employed in the Investigations of chapter v., Part i.

This set of bars was treated somewhat differently from the former—three of them, after being fully magnetized, being slightly reduced by being placed near the test bar, with similar poles contiguous, at the distance of two inches above it.

The effect of this is shewn in the annexed table by the differences, in the case of the first three bars, betwixt the powers in columns III. and IV. In the case of the last three bars (one of them being compound), where the powers in column IV. are marked with an asterisk, no reduction of the maximum powers was made. The powers of the bars, when first placed in the cabinet, October 22, 1841, are given in column IV.; their respective powers, when examined March 21, 1842, in column V.; and their powers when again tried, March 27, 1843, in column VI. The position in which the bars lay was in the direction of the magnetic meridian; the marked ends towards the north.

Mark of Bars.	Weight in Grains.	Maximum Powers.	Powers as placed in Cabinet, Oct. 22, 1841.	Powers as tried Mar. 21, 1842.	Powers as tried Mar.27, 1843.	Loss per Cent.		
						In 5 Months.	In 17 Months.	In the intermediate 12 Months.
I.	II.	III.	IV.	V.	VI.	VII.	VIII.	IX.
H. ii.	1404	32·37	32·8	31·50	31·3	1·1	4·1	3·0
T. ii.	1535	28·55	27·40	27·26	27·23	0·9	1·1	0·2
E. ii.	1518	20·56	19·37	19·22	19·23	1·0	1·1	0·1
T. iii.	710	22·57	22·57*	22·25	22·25	2·4	2·4	0·0
E.iii.iv	984	21·46	21·46*	21·10	20·49	3·0	4·8	1·8
S. i.	3650	18·52	18·52*	18·45	18·29	1·1	2·4	1·3
Mean Results -		—	——	——		1·6	2·7	1·1

Comparing the tangents of the deviations

given in columns iv., v., and vi.; we obtain the loss per cent. in each bar after an interval of five months, as given in column vii.; of seventeen months, as given in column viii.; and in the interval of the intermediate twelve months as given in column ix. Three of the bars, it will be observed, suffered scarcely a sensible loss,— the mean loss being only one-tenth per cent. The greatest loss, it is remarkable, was found in the case of the hardest bar, н ii. But this seems so contrary to analogy, that a suspicion necessarily arises that it must have received some accidental deterioration whilst under trial.

Here we find the average loss, in the first five months, to be 1·6 per cent., and in the intermediate twelve months 1·1 per cent.; total 2·7 per cent. If we compare the bars which were *slightly* reduced by proximity to the test-bar, with those placed in the cabinet at their maximum energy, we find an average loss of 2·1 per cent. in the former, and of 3·2 per cent. in the latter—a result which was naturally anticipated. Were we, indeed, to omit the apparently defective experiment of bar н ii., we should have a mean deterioration in the unreduced bars of 3·2 per cent. in seventeen months, and of the reduced bars of 1·1 per cent.; and a mean deterioration of the same, in the last twelve months, of 1·0 and 0·1, respectively. It will be observed that the proportional loss of power does not, in this case, seem to be so materially in relation to the degree of hardness as might have been anticipated; and that magnets of moderate powers, of any measure of hardness, if placed properly in the magnetic meridian, and kept at a distance from all other magnetic influences, are capable of retaining their powers, with little or almost no loss, for long periods of time.

The bars employed in these latter experiments were re-magnetized to their highest capabilities immediately after their several powers, as given in column vi. of the preceding table, had been determined. Their powers having been again tried, they were placed, as before, in a horizontal position in the cabinet, with their marked ends towards the north. On the 27th of October, 1843, an interval of exactly seven months, they were again examined. No perceptible loss of power, on the average of the series, could be detected; on the contrary, some of the bars seemed to have received a slight augmentation of power by the favourable action of the earth's magnetism. The mean powers of five bars remained the same as at first. The compass, however, having in the mean time undergone repairs, and the agate in the central cap having been changed, it was impossible to determine whether or not there might be some slight alteration in the centring of the needle sufficient to effect, in a very minute degree, the indications of the instrument.

The experiments, I may here observe, relating to the property of steel for sustaining the magnetic energy, are matters of very great delicacy of management, because of the extreme susceptibility of that property under any extraneous magnetic influence. For not only may an instantaneous approach of one magnet within the sphere of influence of another magnet alter the power of the former, or of both; but even a change in the position of a highly magnetized bar from the horizontal to the vertical, may produce a very perceptible alteration in the deviating power of such bar.

RESULTS.

THOUGH this part of my Investigations has not been carried to the extent which I could have desired, and, though the results are not in every respect accordant with presumed analogies; yet so much appears satisfactory as to justify, I think. these general conclusions:—

1. That the degree of *retentiveness* of magnets is directly as the hardness, and inversely as the energy.

Taking all the causes of deterioration into account, such as the action of terrestrial or other magnetism in unfavourable directions, this proposition will be found a general law; though under circumstances of favourable position, in regard to the direction of the earth's magnetism, the effect of hardness, on the one side, or of great energy, on the other, may not, in moderate intervals of time, be very perceptible.

The difference in retentiveness observable betwixt the softer and harder series of plates described at the conclusion of § 1, is corroborative of the proposition under consideration—the loss sustained by the *softer* series in 64 days being 9·88 per cent., and by the *harder* set only 5·82.

2. That the loss of energy by time, in unprotected magnets, whether single or compound, is much more considerable at first than subsequently.

> In the case of the two compound series of smaller plates described at page 354, we find that the harder lost two-thirds as much power the first day, as it subsequently did in 64 days.

> And in the case of the several six-inch bars, the effects of time on the magnetism of which are shewn in the table at page 357, nearly one-half more deterioration occurred in the first five months than in the subsequent twelve months.

3. That *combinations* of magnetized plates of tempered or hard steel, appear to be fully as favourable for the retention of the magnetic energy as single magnets of similar mass, and probably more retentive than single bars of corresponding power,—especially if the plates in combination be separated to a little distance from each other so as to attenuate the *intensity* of the energy in the compound magnet.

> This result is deduced from the considerations drawn from a comparison of the tables at page 352 and page 357.

4. That in magnets of moderate powers, though fully magnetized, if kept clear of deteriorating influences and placed in the direction that would be assumed by the suspended compass needle, the power is very enduring; and but little, if at all, less so, in *soft* than in *hard* magnets.

In the first series of experiments described in sect. 2,
consisting chiefly of plates adapted, as to weight, for
compass needles, the loss of power when laid in the
magnetic meridian was scarcely 2 per cent. in the first
two months, which loss was not perceptibly increased
during the subsequent period of a year.

Again, in the second series of experiments of sect. 2, the
same result is partly verified. And in a subsequent
experiment, not included in the tables, no loss of power
could be at all discerned in a period of seven months,
even after the bars had been re-magnetized to their
highest capabilities; nor was any perceptible difference
observed, in this case, betwixt the sustaining power of
the soft and that of the hard bars.

Hardness, therefore, becomes most essentially important
for resisting the effects of *unfavourable magnetic action*,
whether arising from the earth's magnetism operating
against that of the magnet, or from unfavourable contact with, or proximity to, any other magnet. In
another respect, also, as has been repeatedly shewn,
hardness is of essential moment; that is, for enabling
large magnets to sustain the violence belonging to the
condition of powerful combinations, or massive bars.

In the experiments with the two energetic compound bars,
of different degrees of hardness, described at page 354,
the harder bar exhibited an advantage over the other in
its relative tenaciousness; but in this case the magnets
were very powerful ones, and the position in which
they were laid by (east and west), had not the advantage of terrestrial magnetism for assisting in retaining
the power.

5. That when the maximum power of a
magnet is slightly reduced by unfavourable

364   MAGNETICAL INVESTIGATIONS.

proximity to another magnet, the resulting
energy is still less influenced by time.

This is a result which had been fully anticipated, and is
well illustrated by the experiments exhibited in the table
at page 357. For here we find (excluding the first bar
in the table as an anomalous case, and apparently
defective in accuracy) that whilst the unreduced bars
lost, on an average, 3 2 per cent. in seventeen months,
the bars that had been slightly reduced lost but 1·1
per cent. of their power; and that these reduced ones,
in the last twelve months, sustained scarcely any sensible
loss.

END OF PART II.

London : Printed by Manning and Mason, Ivy-lane St. Paul's.

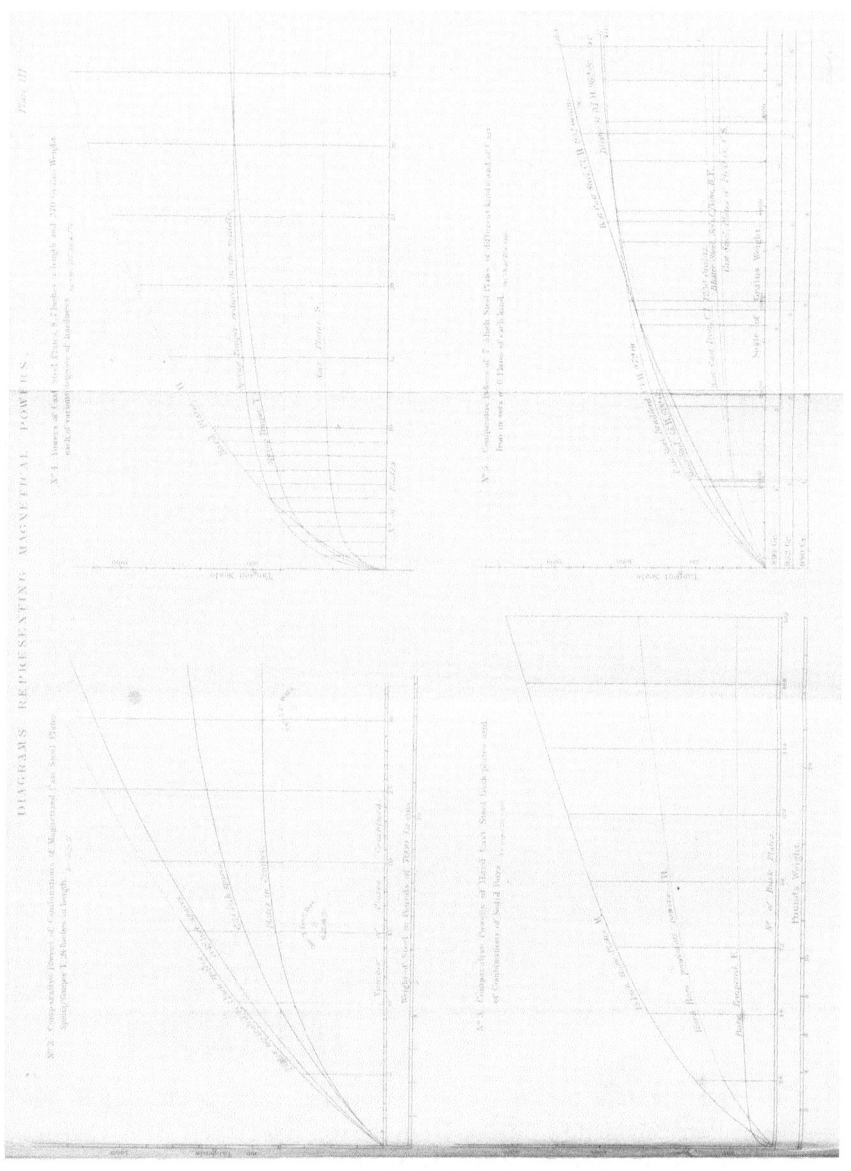

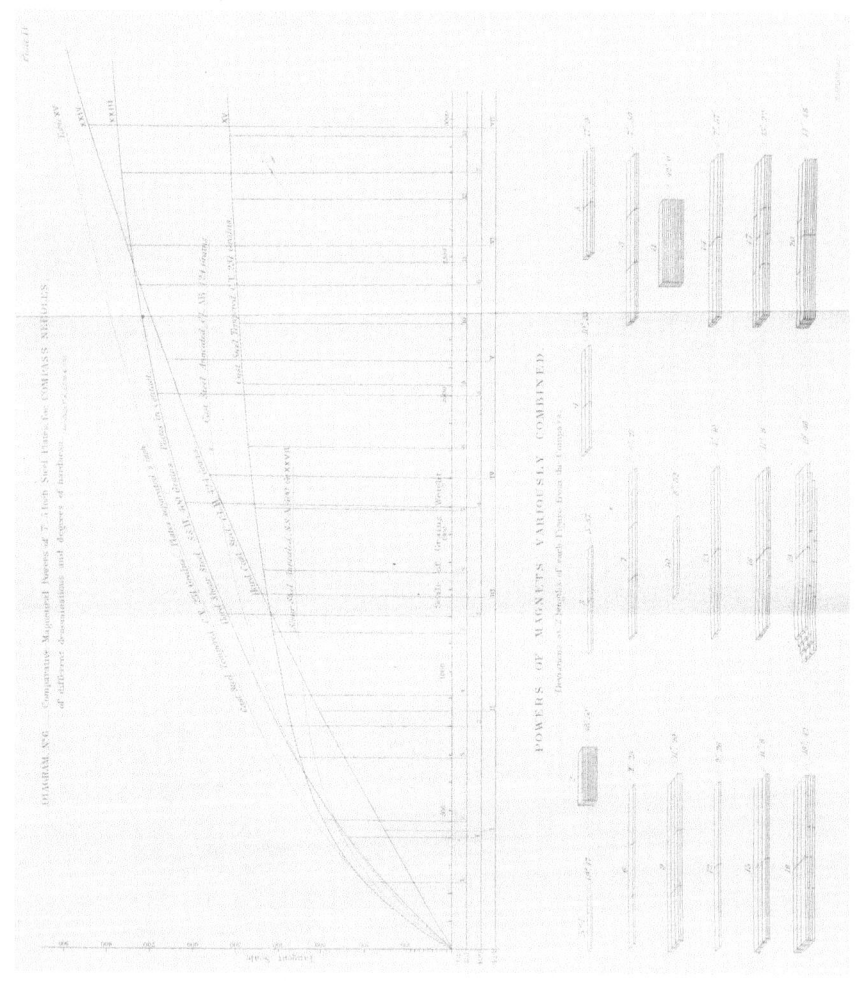

The material originally positioned here is too large for reproduction in this reissue. A PDF can be downloaded from the web address given on page iv of this book, by clicking on 'Resources Available'.